Creating a Culture of Predictable Outcomes

Creating a Culture of Predictable Outcomes demonstrates the importance of creating cultures in the design and construction industries grounded in sophisticated-caring leadership, high-performing collaborative teams, and master-level decision-making discipline, informed by values, to finally address massive inefficiencies, waste, and unpredictability.

Barbara White Bryson offers specific guidance to industry stakeholders to succeed in achieving project-related predictable outcomes by focusing on culture rather than process. This includes selecting the right team members by hiring and firing bravely, valuing psychological safety, leading with values, practicing respect and transparency, fostering empowerment to make decisions at the right level at the right time, and more.

This book is a must-read for design and construction professionals who want to finally understand how to set goals and meet those goals for their clients as well as for their teams.

Barbara White Bryson, Ed.D, FAIA, MBA, is Associate Dean for Research and Director of the Drachman Institute at the College of Architecture, Planning, and Landscape Architecture at the University of Arizona where she also served as Vice President of Strategic Planning and Analysis. At Rice University, Dr. Bryson led an aggressive building program for over a decade as Associate Vice President of Facilities Engineering and Planning. In 2009, she wrote *The Owner's Dilemma: Driving Success and Innovation in the Design and Construction Industry* with Canan Yetmen.

In her straightforward manner, Barbara White Bryson speaks directly to the fundamental misses inhibiting the built environment industry from firing on all cylinders. A blend of in-depth domain knowledge and relatable story-telling, *Creating a Culture of Predictable Outcomes* will stand as foundational to the professions for years to come.

– David G. Gilmore, President & CEO, DesignIntelligence

Working with Barbara on the Rice Bioscience Research Collaborative was one of the best project experiences our team at SOM has had. Her real-world experiences, including senior academic and strategic leadership, as well as being an architect with an MBA and doctorate in higher educa-tion, make Barbara uniquely adept at describing comprehensive and highly effective strategies with analogous references that are clear, engaging, and memorable.

– Carrie Byles, FAIA, LEED BD+C, Partner SOM

"Predictability is power." There are no truer words spoken than this premise from Barbara White Bryson's illuminating and strategic book, particu-larly in a time of chaos and mismanagement. Barbara's book is an immi-nently useful read for firm leaders at any phase in their careers. Universally, business, design, and engineering schools will use this book to demonstrate predictability as the active decision-making model we need.

– James Timberlake, FAIA, Architect/Partner,
Kieran Timberlake, Philadelphia

Creating a Culture of Predictable Outcomes packs a powerful message while forging a new path for the industry, one that is more nimble, adaptable, and resilient. Bryson's list outlining twenty disruptions that will impact our future is alone worth the price of the book. As we've all witnessed, unex-pected things happen—this book is here just in time to help!

– Randy Deutsch, Author of *Superusers* (Routledge, 2019)

Creating a Culture of Predictable Outcomes

How Leadership, Collaboration, and Decision-Making Drive Architecture and Construction

BARBARA WHITE BRYSON

Routledge
Taylor & Francis Group

NEW YORK AND LONDON

First published 2021
by Routledge
52 Vanderbilt Avenue, New York, NY 10017

and by Routledge
2 Park Square, Milton Park, Abingdon, Oxon OX14 4RN

Routledge is an imprint of the Taylor & Francis Group, an informa business

© 2021 Taylor & Francis

Library of Congress Cataloging-in-Publication Data
Names: Bryson, Barbara White, author.
Title: Creating a culture of predictable outcomes : how leadership, collaboration,
and decision-making drive architecture and construction / Barbara White Bryson.
Description: First edition. | New York, NY : Routledge, 2021. |
Includes bibliographical references and index.
Identifiers: LCCN 2020034120 (print) | LCCN 2020034121 (ebook) |
ISBN 9780367894382 (hardback) | ISBN 9780367894375 (paperback) |
ISBN 9780367894399 (ebook)
Subjects: LCSH: Building–Superintendence. |
Design-build process (Construction industry) |
Architectural practice. | Corporate culture.
Classification: LCC TH438 .B745 2021 (print) |
LCC TH438 (ebook) | DDC 624.068/4–dc23
LC record available at https://lccn.loc.gov/2020034120
LC ebook record available at https://lccn.loc.gov/2020034121

ISBN: 978-0-367-89438-2 (hbk)
ISBN: 978-0-367-89437-5 (pbk)
ISBN: 978-0-367-89439-9 (ebk)

Typeset in Dante and Avenir
by Newgen Pulishing UK

To my son, Ian Stowe. By watching him embrace wondrous, entrepreneurial adventures in architecture and construction, I have faith his generation will lead us to a remarkable future.

Contents

List of Figures

Graphics for chapter figures by Emily Chang unless otherwise noted.

Cartoons by Barbara White Bryson

Preface

Oh, what a superpower it would be—to predict the future, to know exactly what path to travel, to know the perfect solution to every problem, and to understand without doubt how to prepare. Unfortunately, knowing the future or predicting the future is not possible. The only thing you can predict about the future is it will surprise you. This past year, 2020, has certainly surprised me. However, even without the ability to be prescient, I believe we can predict outcomes on design and construction projects with reasonable certainty, barring natural disasters, pandemics, or war, but controlling those outcomes takes effort and thoughtfulness.

I know something else about the future. My time to impact my profession is coming to an end. As futurist Ray Kurzweil has said, "Death gives meaning to our lives. It gives importance and value to time. Time would become meaningless if there were too much of it."[1] I am in no danger of leaving the planet yet, but, like many Baby Boomers, I am nearing retirement age. Even Boomers don't last forever. I still want to be engaged and useful. However, it is apparent to me it is time to pass the baton.

I watch so many of my contemporaries holding on for dear life, refusing to understand that they may no longer have the best ideas, the most energy, or the wisest counsel. But, the world is moving on, and so are the industries of design and construction. The paradigms have shifted, and the Boomers did not get it done. We did not manage to make these industries significantly better than when we started. We barely changed them at all. There have been so many lost opportunities—so many moments when we might have leaped. Still, we kept allowing ourselves to be pulled back into our comfort zone, the zone we understood best, the adversarial zone, the build the building "one stick, one brick at a time" zone.

So, it is time to let it go, Baby Boomers. Not that we really have any choice. As our bodies and minds fail, there are younger, stronger, and more energetic professionals ready to take our place. Unfortunately, through an accident of birth rates and a great recession, there are too few GenX leaders to take over for us Boomers, an excellent excuse for holding on longer. Now, however, that we Boomers are at the end of our run, we are looking more and more to Millennials to provide project and studio leadership. I have seen the panic on my contemporaries' faces as they tell me that these thirty-somethings aren't ready to take the reins. Yet, I know that each of these great firm leaders was given their first amazing opportunity when they were about that same age. These young leaders are more than ready, and, hopefully, they are prepared to do things their way, a different way.

As I take stock of our industries, I am still frustrated. Our industry remains fragmented and adversarial. We are weighed down by a lack of foundational information for decision-making. We are still victims of the vagaries of the economy and still building many buildings one element at a time. Technology and the economy present many opportunities for disruption, but it is not clear that the traditional professionals of the industry will ever leverage these opportunities. Change is coming. That is inevitable. I want to believe our emerging professionals will be ready for this change as they pull the reins from the reluctant hands of the Boomers. However, unless these few GenXers and numerous Millennials take a new approach to the design and construction industries, they will fail just as surely as we Boomers have.

I wrote this book because "time has value to me," and in the time I have left to make an impact, I hope that we can, together, shake things up and forge a new path. I hope that young professionals can create new environments in which they will design and build in radically different ways. I hope they will not be satisfied with incremental change, but energetically and bravely seek profound change. I challenge you to disrupt these industries from the inside. If that does not happen, disruption will arrive uninvited. To all generations of designers and builders, I call you to action. Start by building a Culture of Predictable Outcomes. These ideas are my gift to you as I let go.

- *Barbara White Bryson*

Note

1 "Take death for example. A great deal of effort...," Quotetab, June 16, 2020, www.quotetab.com/quote/by-ray-kurzweil/take-death-for-example-a-great-deal-of-our-effort-goes-into-avoiding-it-we-make.

Acknowledgments

I am so grateful for the love, support, humor, and cooking of my husband, Jeff.

Thank you to the many friends and colleagues that have supported, listened, debated, and read, including Susann Glenn, J.D. Schramm, James Timberlake, Randy Deutsch, Carrie Byles, Emre Ozcan, Mike McCormick, Dave Gilmore, Bob Fisher, and Nancy Pollock-Ellwand Thanks to all those that lent their stories to chapter supplements and other examples in this book. The ideas are richer because of your help. Thank you to my teacher, Bob Zemsky. I am forever your grateful student.

Special thank you to Emily Chang for providing graphics within these chapters!

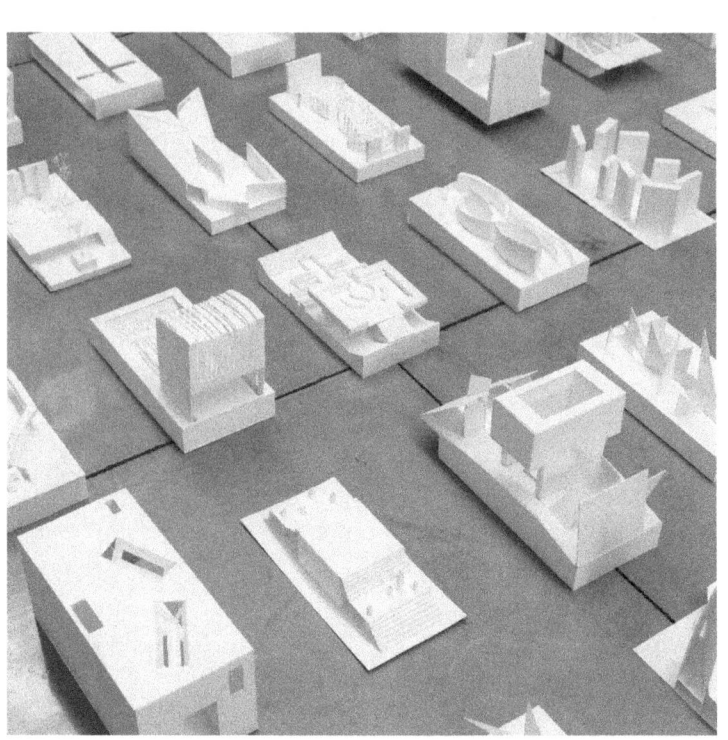

Predictability Is Power 1

Unpredictable

What would you do if you could predict the future? Would you bet on the horses, invest in a tech company? Would you befriend a future movie star, or avoid an accident on the freeway? Would you avoid that lettuce at the salad bar, save a life, or marry the perfect mate? Or, would you pick the winner of the World Series three years in the future just as Ben Reiter of *Sports Illustrated* famously did in 2014? As a baseball fan, that might be my choice.

You see, predictability is power. When trains run on predictable schedules, we make it to meetings or the airport on time. When a stoplight turns from green to yellow, then to red, we know how to react. We stop and then start in a civilized fashion. When Starbucks gets your "Tall, Non-Fat Latte with Caramel Drizzle" perfect every time, even though they never spell your name the same way twice, all is right with the world.

Predictability is power. If those trains do not run on time, you would have to leave for appointments or to catch airplanes hours earlier. You might choose cars instead, further impacting traffic, air quality, accident rates, insurance, and general frustration. If your caffeine of choice is not delivered perfectly every time, well, the world wouldn't change very much, but an amazing business would not exist, a business built on predictable outcomes.

Predictability is not something we see much of in the design and construction industries. In August 2017, *The Economist* reported more than 90 percent of the world's infrastructure projects were not meeting schedule goals, and more than 60 percent of the UK's building projects were not meeting budget goals.[1] In 2016, a McKinsey report documented "large projects across asset classes typically take 20 percent longer to finish than scheduled and are up to 80 percent over budget."[2]

To experienced professionals in the design and construction industries, these reports are not surprising or unique. In 2010, I wrote in *The Owner's Dilemma: Driving Success in the Design and Construction Industry* that a Google search with the phrase "buildings over budget" pulled more than 48 million hits, including a hospital over budget by a million dollars and an Olympic village over budget by a $100 million.[3] A similar search in 2018 unearthed a Chinese dam with a budget that quadrupled from $8.5 billion to $37 billion, a BBC headquarters with an overrun of 110 million pounds, and a 35,500 square foot, $13 million dream home that cost $20 million before it was complete. These data bolster the argument from the article in *The Economist* noted earlier, which also distressingly reported on the ill-fated Berlin Brandenburg airport. That project, self-described as a "shit-show," was, at the time, six times over budget.[4]

Take it! You should have told me you're a contractor. YOUR future is UNPREDICTABLE!

The design and construction industries are riddled with these unpredictable outcomes, and, unfortunately, most industry professionals appear powerless when asked to dependably deliver projects within budget and schedule goals. As a result, achieving consistently predictable outcomes has become the holy grail of the design and construction industries. Predictably attaining quality, schedule, and budget goals set at the start of a project should be a minimum standard of performance for any industry. Yet, predictable attainment of goals seems to be an impossible dream in the design and construction industries, known as the least efficient and productive in the world.

This lack of predictability is the most glaring evidence of our industries' inefficiency, which remains the lowest of any major industry type in the world. It is, however, not the only evidence. Amy Edmondson and Susan Reynolds listed a number of unfortunate facts from multiple sources, including the Modular Building Institute, in their book, *Building the Future*, a book about complex super teams. The authors noted,

> Research on construction industry efficiency reveals "25 to 50 percent waste in coordinating labor and in managing, moving, and installing materials […]; losses of $15.6 billion per year due to the lack of interoperability […]; and transactional costs of $4 billion to $12 billion per year to resolve disputes and claims."

Edmondson and Reynolds added, "Studies suggest that up to 75 percent of construction activities **add no value**."[5]

Where other industries have been able to restructure and reinvent, the design and construction industries continue to slog along, refusing to address their fragmentation, supply chain weaknesses, labor challenges, and communication barriers. The McKinsey Global Institute reported in February 2017 that "Globally, construction sector labor-productivity growth averaged 1 percent a year over the past two decades, compared with 2.8 percent for the total world economy and 3.6 percent for manufacturing."[6] Despite technology advances, according to McKinsey, productivity continues to disappoint in the construction industry, yet, if we were able to match the efficiency of the rest of the economy (manufacturing, retail, and agriculture), we could add $1.6 trillion annually to the world economy. This amount would "meet about half of the world's annual infrastructure needs or boost global GDP by 2%."[7] That number reflects impact and opportunity. There is no doubt that Katerra, Convene, UniSpace, and other bright nontraditional companies have their eyes on precisely that target and those opportunities.

When I wrote *The Owner's Dilemma* with Canan Yetmen in 2010, I was already aware positive outcomes in design and construction were related to the predictability of those outcomes. You see, owners need to understand

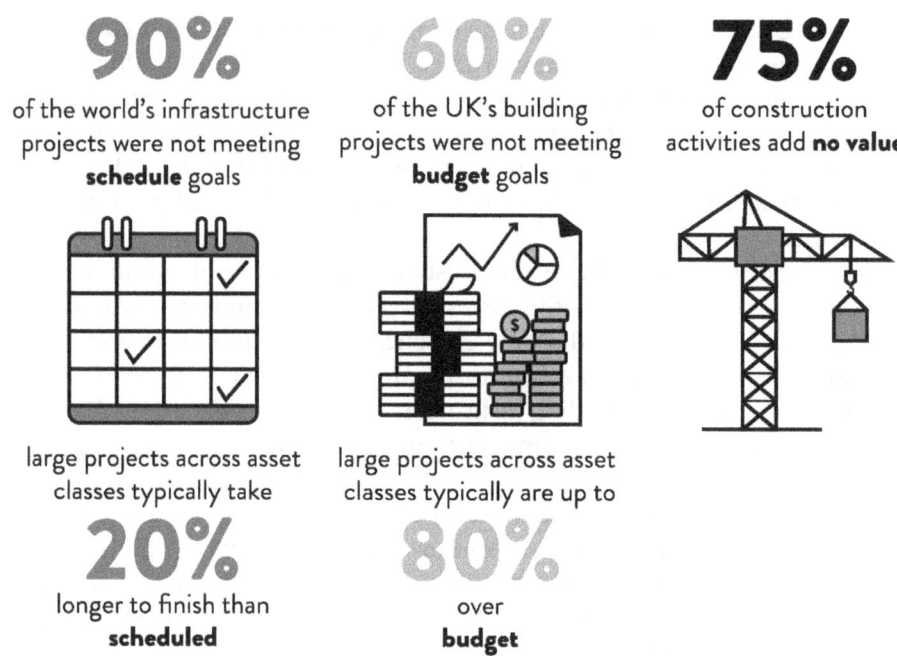

Figure 1.1 Data demonstrates lack of efficiency of the design and construction industries

what they are getting as far as cost, schedule, and quality at the beginning of each project. The results don't always need to be the cheapest or the quickest, but sometimes they do. There is no doubt, though, owners always know precisely what product—what building—they need for their business, and if they are not going to get that product, they need to know as soon as possible. That product—that building—impacts their business plan. The need to understand what you are getting as an owner is consistent for universities, hospitals, corporate owners, developers, and other business owners, large and small. COVID-19 has brought this priority into even clearer focus as business plans have critically tightened.

Unfortunately, for most industry professionals, the search for stability and control in the industry starts with the selection of process. By process, I mean delivery processes such as Design-Bid-Build, Integrated Project Delivery (IPD), Design-Build, or GMax. Many in the design and construction industries still mistakenly believe in the magic of selecting the perfect delivery process. We are obsessed with this choice. Inevitably, the very first question I am asked when I speak at industry events is, "What is the best delivery process?" Many industry self-help books spend most of their pages analyzing delivery processes and related contracts that only incrementally improve the industry. Some books are dedicated entirely to a single process like design-build or IPD. This obsession is tantamount to Nero fiddling while Rome is burning.

It's Not About the Process

I will not argue that some processes do not lend themselves to greater predictability than others through increased communication and lowered risk. I am not saying that process is not essential in our industries. However, it is well documented that a **lousy team can screw up a great process,** and, conversely, a **great team can overcome even the worst processes and contracts**. I've learned this lesson repeatedly during my career, as many professionals have. I have seen great teams deliver Design-Bid-Build projects successfully, and terrible teams screw up the best collaborative IPD contract. Certainly, good processes can help support a great team, a team with values, and a team actively collaborating. Still, no process can ever overcome the deficiencies of a lousy team or an owner/client that does not take their responsibilities seriously.

Something that significant and progressive architectural firms of the world like KieranTimberlake know, daring and innovative companies of the world like the Broad Company know, as well as forward-looking enterprises like McKinstry know, and emerging property technology companies know, is that **a great team can reinvent and create transformational processes**. I wrote an article a couple of years ago called "The Future of Architects: Irrelevance or Extinction" as a call to action.[8] I worry our resistance to change in our profession will be our undoing. It is important to recognize that our professions are not sacred and our places in these industries are not guaranteed. We in the design and construction industries must wake up and understand that incremental change is no longer sufficient.

Do not confuse predictability with the ordinary in the design and construction industries. Predictability in our industries would be extraordinary and remarkably valuable. Time after time, owners and project teams are disappointed by delivery processes and deeply confused by the number of ways projects can go wrong. This confusion stems from the fact that predictable outcomes for projects are primarily a **product of culture, not process.**

When I read General Stanley McChrystal's book *Team of Teams*[9], about creating new kinds of resilient teams to respond to a new type of warfare, I had a critical moment of insight. I realized many people in our industries approach designing and building projects as if they are going to war with the members of their teams. These adversarial relationships have long undermined our ability to fight the real enemies of our industries, the fragmentation, the complex challenges, the unexpected problems, the changing conditions, the lack of knowledge, the lack of decisions, and the lack of communication. We have all experienced this destructive dynamic emerging from the overwhelming desire and responsibility to protect our businesses or our interests above what is right for the entire team and the entire project. Owners do this all the time, treating the project team as the enemy. Nothing good comes out of war. As Pulitzer Prize-winning author Barbara Tuchman said, "War is the unfolding of miscalculations."[10] We have enough confusion and miscalculation in our industries with which to grapple without creating more. These are the teams that destroy processes, even great processes. Mark Linenberger from Linbeck observed IPD team members having trouble building trust on Cook Children's Medical Center because it's "easy to become protective of your interests and forget that in IPD the team wins or loses together."[11] When we join a team, we must be fully committed to that team and to creating a culture that leverages processes and produces great outcomes.

A Culture of Predictable Outcomes Drives a Project to a Set of Clear Goals and Achieves Those Goals

The Owner's Dilemma was as much a leadership book as a project management book. However, it also addresses two other key points impacting project success: collaboration and decision-making. It was not until years later and additional research I understood that predictable outcomes are a product of a working culture created by the convergence of excellence built into each of these three areas, Leadership, Collaboration, and Decision-Making. More specifically, a Culture of Predictable Outcomes is a culture built from sophisticated-caring **leadership**, high-performing **collaborative teams**, and master-level **decision-making discipline**, all supported by skills, processes, and tools (See Figure 1.2). If any of these elements are missing, predictable outcomes will be compromised, even corrupted. With these elements, predictable outcomes, barring natural disasters, are guaranteed. I found this pattern repeated again and again in a variety of successful teams, whether it relates to a campus construction project, a Chilean mine disaster, or the Astros 2017 World Championship Baseball organization.

A Parable

On June 30, 2014, the only thing that seemed predictable about the Houston Astros baseball team was its losing record. Ben Reiter, sportswriter and,

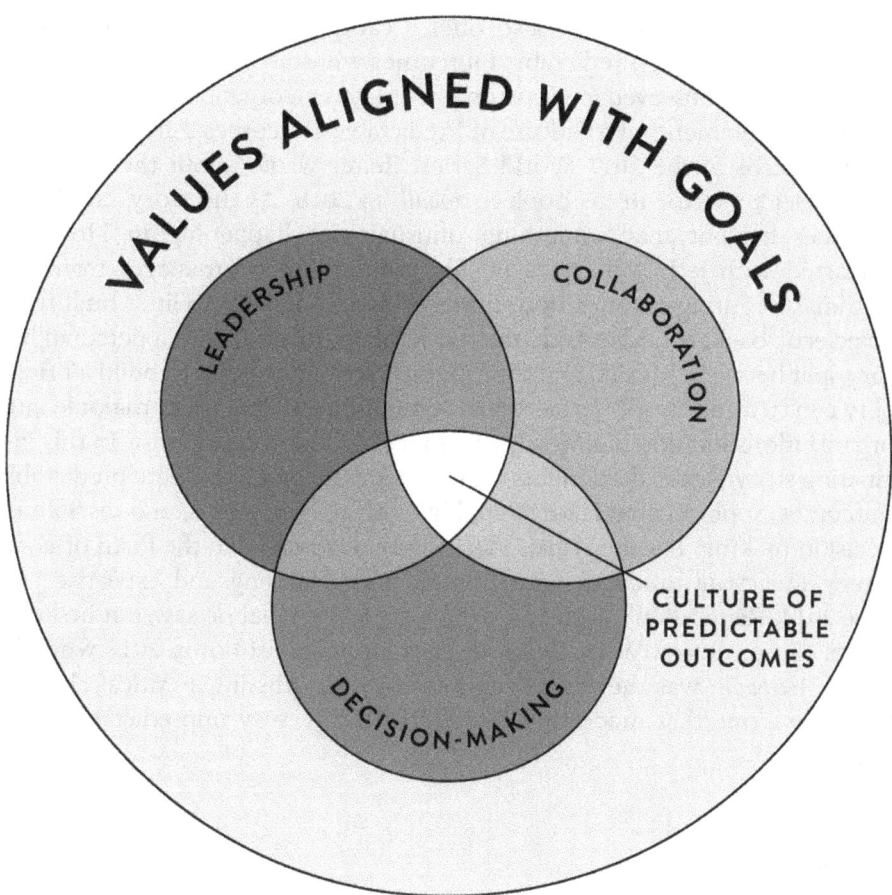

Figure 1.2 Model for creating a culture of predictable outcomes

later, author of *Astroball,* described the team as "ludicrously bad."[12] In 2017, Ben reflected, "Since the start of the 2011 season, the Astros had lost nearly twice as many games, 358, as they had won, 187."[13] Regardless of the Astros' record, Ben went back to New York in June 2014 after meeting with the Astros organization and wrote a cover story for *Sports Illustrated* predicting the loser Astros would be World Series champions just three years later in 2017. The rest of the world was stunned when the Astros won the World Series championship exactly three years later, and when George Springer, the man featured on the 2014 cover of *Sports Illustrated*, appeared again on the 2017 cover as the World Series MVP.

What did Ben Reiter see when he visited the Astros that June? What does Reiter's prediction have to do with the design and construction industries?

How does the Astros' story also offer a cautionary tale? Our journey to building a Culture of Predictable Outcomes will start with these questions.

What Reiter observed in Houston in 2014 was an organization developing some of the elements of a Culture of Predictable Outcomes. After the success of the Astros in the 2017 World Series, Reiter wrote about their journey from loser to victor in his book *Astroball*, in 2018. As the story unfolds, it becomes evident that something unusual was happening in Houston. It started with Jeff Luhnow, a freshly minted general manager from the Cardinals scouting organization. Luhnow had leadership abilities built from an eclectic background outside the world of sports and, with a personal history split between Mexico and the United States, "an ability to build a bridge between two cultures."[14] Jeff also had something else that other major league organizations did not: Sig Mejal, a former NASA rocket scientist. In this fascinating story, Reiter documents critical elements that can create predictable outcomes, sophisticated leadership, high-performing teams, and disciplined decision-making. For the Astros, decision-making came in the form of combining *Moneyball* practices with human understanding and expertise.[15] In June 2014, Reiter didn't quite know how to define what he saw, but he knew it was extraordinary. What Reiter did not know in 2014 or in 2018 when he wrote *Astroball*, was the Astros organization was missing a critical element to success, one that made their long-term future very unpredictable, even scandalous.

The Elements

The three critical elements of a Culture of Predictable Outcomes are sophisticated-caring **leadership**, high-performing **collaborative teams**, and master-level **decision-making discipline**. The formula is that simple, although specific qualities of each element are important. These qualities will be outlined in Chapters 2, 3, and 4. Although daunting, these elements are straightforward and achievable on any project. So why is creating a Culture of Predictable Outcomes so difficult to attain?

There are four reasons for this failure grounded significantly in a lack of business training for professionals in various parts of our industry. Because business skills are not generally part of an architecture or engineering education, these professionals often default to tactics rather than strategies. These tactical approaches have created barriers to success:

1) The first barrier to success is the **design and construction industries' obsession with delivery processes**. Many in our industry believe processes drive collaborative behavior, yet there is ample evidence demonstrating this is not the case. Collaborative team behavior is supported by process not driven by process.

   In addition, delivery efficiency has been confused with procurement efficiency and, as a result, often creates the opposite outcome. There was a time where the architect was also the master builder. The entire team worked together to meet the project goals. Once specialists emerged and architects created contract drawings to communicate design intent to general contractors, the procurement process changed. Boards and legislatures believed they had a tool for competitive procurement.

   This history is well documented in *Program Management 2.0* by Thomsen and Sanders, "Governments at all levels mandated competitive bidding with legislation and procurement regulations. Now that there were drawings and specifications that described the project and owners could ask several builders for a price, they could award the job to the lowest bidder. That was particularly attractive to public owners who wanted to demonstrate an objective use of public money. Although it was far from a universal fact, it gave the appearance of buying the project for the fewest tax dollars. Conventional wisdom was that design-bid-build gave everyone an equal opportunity to do business on the public dollar and the process precluded graft."[16]

   Unfortunately, this new procurement process had the unintended consequence of creating adversarial relationships. It also led to a dynamic where many contractors felt forced to bid low and use change orders to make a profit. So began the myth that delivery processes are the primary source of success and failure for the industry. Unfortunately, change in delivery processes represents only incremental steps, distracting from the other enormous challenges of the industry, the fragmentation, one-brick-at-a-time construction, start-from-scratch-every-time design, resource-consuming end products, and incredibly expensive and inefficient traditional practices. No delivery process by itself, Design-Build, Progressive Design-Build, G-Max, CMAR, or IPD, effectively addresses these issues.

2) The second barrier to success is a lack of understanding that **all three elements of Creating a Culture of Predictable Outcomes are critical to success**. Sophisticated-caring leadership, high-performing collaborative teams, and master-level decision-making are essential and

synergistic elements to creating a culture of predictability and success. Yet, many organizations focus only on one or two of these elements with disappointing results:

- **Leadership and Collaboration alone,** not supported with disciplined decision-making, is unproductive—Project teams focused solely on leadership and collaboration are doomed to frustration without disciplined decision-making. Unless leaders and team members understand how and when decisions will be made along with what information will be required to make those decisions, projects will stall under decision weight.

- **Leadership and decision-making alone**, without collaboration, destroys teams—Project teams relying on leadership and decision-making without focusing on collaboration undermine their teams, ultimately destroying team effectiveness, causing mistakes and delays. Leaders making decisions without teams will not get team buy-in to decisions and may find those decisions undermined. Teams are needed to provide decision-ready information; therefore, decisions made without team involvement will be less informed and highly suspect.

- **Decision-making and Collaboration alone**, without leadership, will simply break down—Project teams that rely on decision-making and collaboration are missing critical leadership support providing a clear path to success. Without sophisticated-caring leadership, even the most collaborative teams will stray off course, and some critical decisions, decisions owned only by leaders, will not be made. In the end, teams and projects without leadership will break down.

3) A third barrier to pursuing a Culture of Predictable Outcomes is our traditional **need for control in the design and construction industries**. We tend to be incredibly hierarchal and spend a great deal of time worrying about who is in charge. Is it the architect, owner, contractor? This interest in control is very old-fashioned and is counterproductive to creating great, nimble teams. Control slows down knowledge-building, innovation, and decision-making at the very moment that we need more speed. Stanley McChrystal spoke about a similar challenge in his book *Team of Teams,*

> "In the course of this fight against terrorism, we had to unlearn a great deal of what we thought we knew about how war—and the world—worked. We had to tear down familiar organizational structures and

rebuild them along completely different lines, swapping our sturdy architecture for organic fluidity, because it was the only way to confront a rising tide of complex threats."[17]

Holding on to control will kill innovative cultures.

4) The last barrier to success in achieving a Culture of Predictable Outcomes is not necessarily learned in business classes. However, this element is often the missing ingredient in many leadership, collaborative, and decision-making cultures, ensuring they will fail. The missing ingredient is **"Values Aligned with Goals."** This was the missing element for the Astros in 2017 when they began their ill-fated sign-stealing activities alongside their run to the 2017 World Series. It was the missing ingredient as the Astros manager, Hinch, found himself sabotaging the illicit monitor in the dugout but never confronting the players,[18] and it was the missing component as the players took control of the Astros' future while leaders turned a blind eye.

"Values aligned with goals" is an overarching ingredient in creating a Culture of Predictable Outcomes (See Figure 1.2). Let's be clear, the Astros team had a set of values in 2017, but those values were aligned with winning at all costs. They did not align with maintaining credibility for the club's leaders, building a future for the club's players, or even maintaining the championship they won in 2017, which might have been at risk. General Manager Luhnow stated that he was not aware of the activities of his organization. Yet, the evidence clearly showed that the baseball operations employees and the players consistently believed that Luhnow, like Hinch, endorsed their actions through his silence and inattention. Values must be articulated.

In the design and construction industries, many firms and companies spend little or no time discussing values and the manifestations of those values. The lack of aligned values to goals undermines our ability to create cultures that operate in a predictable and innovative manner.

What's Ahead

This book is designed to help project teams and other stakeholders in the design and construction industry overcome these barriers by introducing you to each of the elements and underlying qualities of Creating a Culture of Predictable Outcomes.

Chapter 2—In this chapter, **Sophisticated-Caring Leadership** will be carefully defined, addressing both character and responsibilities.

Chapter 3—In this chapter, **High-Performing Collaborative Teams** will be provided with a guide to success.

Chapter 4—In this chapter, **Master-Level Decision-Making Discipline** will be framed and detailed so team members can use these tools to effectively drive outcomes.

Chapter 5—In this chapter, **Aligning Values and Goals** will be introduced as the heart of the Culture of Predictable Outcomes.

The challenges and the opportunities related to **Risk and Contracts** will be discussed in Chapter 6. In Chapter 7, the **Twenty Freight Trains of Disruption** will be a fitting backdrop for considering a culture of innovation. In Chapter 8, the **Case for Research** will be presented (again).

By the time we get to Chapter 9, you will better understand not only how to remove the barriers to Creating a Culture of Predictable Outcomes but how you can **predict a better future for your projects and organization**.

Chapters 2–9 have supplements to provide you an opportunity to consider and discuss how a Culture of Predictable Outcomes applies to real situations in the design and construction industries.

Notes

1 "Efficiency Eludes the Construction Industry," Business section, *The Economist*, August 17, 2017, www.economist.com/business/2017/08/17/efficiency-eludes-the-construction-industry.

2 Rajat Agarwal, Shankar Chandrasekaran, and Mukund Sridhar, *Imagining Construction's Digital Future*, Report from McKinsey Productivity Sciences Center, June 2016, www.mckinsey.com/industries/capital-projects-and-infra-structure/our-insights/imagining-constructions-digital-future, 1.

3 Barbara White Bryson and Canan Yetmen, *The Owner's Dilemma: Driving Success and Innovation in the Design and Construction Industry* (Atlanta: Ostberg Library of Design Management, Greenway Communications LLC, 2010), 58.

4 "Efficiency Eludes the Construction Industry," Business section, *The Economist*, August 17, 2017, www.economist.com/business/2017/08/17/efficiency-eludes-the-construction-industry.

5 Amy C. Edmondson and Susan Salter Reynolds, *Building the Future: Big Teaming for Audacious Innovation* (Oakland, CA: BK/Berrett-Koehler, 2016), 107.

6 Filipe Barbosa, Jonathan Woetzel, Jan Mischke, Maria Joao Ribeirinho, Mukund Sridhar, Matthew Parsons, Nick Bertram, and Stephanie Brown, "Reinventing Construction: A Route to Higher Productivity," McKinsey

Global Institute, February 2017, www.mckinsey.com/~/media/McKinsey/
Industries/Capital%20Projects%20and%20Infrastructure/Our%20Insights/
Reinventing%20construction%20through%20a%20productivity%20revolu-
tion/MGI-Reinventing-Construction-Executive-summary.ashx, 4.

7 Ibid.

8 Barbara Bryson, "Future of Architects: Extinction or Irrelevance," *Design
Intelligence Quarterly*, June 6, 2017, www.di.net/articles/future-architects-
extinction-irrelevance/.

9 Stanley McChrystal, Tantum Collins, and David Silverman, *Team of Teams*
(New York: Penguin Publishing Group. 2015).

10 Joseph Demakis, *The Ultimate Book of Quotations* (Charleston, SC: CreateSpace,
2012), 499.

11 Martin Fischer, Howard Ashcraft, Dean Reed, and Atul Khanzode, *Integrating
Project Delivery* (Hoboken, NJ: Wiley, 2017), 24.

12 Ben Reiter, *Astroball* (New York: Three Rivers Press, 2018), xi.

13 Ben Reiter, "About that Prediction…How the Astros Went From Baseball's Cellar
to the 2017 World Series," *Sports Illustrated*, October 24, 2017, www.si.com/
mlb/2017/10/24/houston-astros-sports-illustrated-world-series-prediction.

14 Ben Reiter, *Astroball* (New York: Three Rivers Press, 2018), 19.

15 Michael Lewis, *Moneyball: The Art of Winning an Unfair Game* (New York: W.W.
Norton, 2004).

16 Chuck Thomsen and Sid Sanders, *Program Management 2.0: Concepts and Strategies
for Managing Building Programs* (rev. ed.) (McLean, VA: CMAA, 2011), 56.

17 Stanley McChrystal, Tantum Collins, and David Silverman, *Team of Teams*
(New York: Penguin Publishing Group, 2015), 20.

18 Evan Drelich, "Hinch, Luhnow Fired as MLB Report Reveals Details of Astros
Sign-Stealing," *The Athletic*, January 13, 2020, https://theathletic.com/1531133/
2020/01/13/hinch-luhnow-fired-as-mlb-report-reveals-details-of-astros-sign-
stealing/.

Sophisticated-Caring Leadership   **2**

Tony DiCicco, head coach of the 1999 US Women's World Cup team, had decided Carla Overbeck was his team leader even if she wasn't the youngest or the fastest player. "I don't care if the woman was in a wheelchair, we would never be successful unless she was on the field," stated DiCicco. "In some ways I have to coach to protect her a little bit, but I made up my mind that we were going to live and die with Carla in the lineup. She is very good on the ball and she's a great leader with great heart. This is a better team when she is in there."

The stakes couldn't be higher as the American team sought their second World Cup title and the first women's World Cup that had captured the world's attention. The final game would be played in the Rose Bowl, packed with more than 90,000 fans. In this stressful environment, DeCicco wasn't the only one with faith in Overbeck. Teammate Kate Sobrero reflected, "[Overbeck] runs the show. She's the most calm. You hear the voice. You get the sense that no matter what happens, we're going to be okay." It was Overbeck who set the standard for "extreme fitness" on the team, a standard that younger players often had a hard time matching. The author of *The Girls of Summer* recounts, "[Overbeck] ran stadium steps and hills and exercised on a stair-stepper machine during the ninth month of pregnancy."

Another teammate, Brandi Chastain, directly experienced Overbeck's effective leadership style after putting a ball into her own goal in a quarter-final match against Germany. "The first person who reached her was Overbeck. 'This game is not over,' Overbeck told her in a reassuring voice. 'There are 85 minutes left. We are going to win this thing. Don't worry about it. Get over it. Let's do it.'" Chastain remembered, 'I've got this look of "Oh my God I've just passed it into my own goal," and she trusts me …

She gave me confidence to go forward. Sure enough, in the second half, I scored a goal, for us this time, to tie the game and we [went on to] win 3-2. She could have easily looked me in the eye and said, "You blew it." But she gave me the courage to move forward and the trust and respect I deserved from all the years of training. For me, that was my best moment of the World Cup.'"

On the day of the final game of the World Cup against China, team captain Carla Overbeck pulled the team together and exhorted, "Norway screwed us in 1995. This one is ours. We deserve it. We are the best. It's ours to go get it. The only thing between us and the trophy is that team. Let's go fuck 'em up."

After more than two hours of scoreless regular play and overtime, Overbeck took the first of five penalty kicks, leading the team to victory on July 10, 1999.[1]

The Leadership Challenge

Leadership in the design and construction industries is uniquely challenging due in part to the industries' fragmentation, economic fluctuations of the industries, and the frustrating inertia of the professions working within these industries. These dynamics build stress and encourage professionals to approach their work as adversaries rather than team members. Many industry leaders approach projects as if they are armed for battle, trying to beat down the fee or increase change orders—creating environments where only winning is important. This destructive approach has long been responsible for reduced project value and has profoundly inhibited innovation in our industries.

Leading with a hammer doesn't work in any industry, but most especially does not work in the design and construction industries. In this fragmented, multidiscipline environment where there is no one with clear authority over the entire project except, possibly, the owner, leading through influence is as important, or more so, as leading through directive. In these industries, where pushing a project to a successful end can be as frustrating as bailing water with a sieve, sophisticated, caring, supportive, influential, inspiring leaders are needed more than ever.

Oh, yes—I hear your arguments back to me right now. I have heard it all. You believe it is your job, in these fragmented, competitive industries, to do the very best for your company or firm. You have the responsibility not to let anyone take advantage of your organization. Therefore, you must be on the defensive. Unfortunately, when you approach any project with a "my firm first" attitude, you will not win. Instead, you will reduce potential value for every project team member, including your own.

Recently, I spoke with a colleague about a struggling theater project on his campus. This premier performance facility had the advantages of a world-class architect, a significant budget, board of trustees support, and an experienced contractor. My colleague and his department were adept in managing complex and highly political projects. Yet, this project was, based on this account, in trouble. As I listened to my colleague complain about the contractor's behavior as well as the lack of responsiveness of the architect, there seemed to be a number of typical frustrating difficulties affecting the project, nothing more. Then my colleague told me a story that demonstrated a deeper concern. (Note: Details have been changed to protect participants.)

As the project entered its final stages of construction, the design approvals of the main lobby interior had stalled. The dean of the drama department and primary client for the project had decided he did not want the university's Design Review Committee (DRC) to see the design because the dean had already shown it to the donor. The donor had already approved the design. The project manager had known about the donor approval and, because the project was so far behind schedule, used the donor approval as an excuse to proceed with construction even though the DRC approval had not been received. The DRC approval is a board requirement. So when the assistant vice president, my colleague, sat in a meeting and realized that the scaffolding was about to come down revealing a completed lobby interior containing the poorly assembled colors of an arch-rival university, without DRC review, he began to panic.

This story is what lack of leadership looks like, and this is the chaos that follows lack of leadership. You can sense there is a leadership problem when a single stakeholder unilaterally redefines the approval processes or when a donor usurps the administrative approval processes. You might also assume leadership issues exist when approvals are happening months later than required and when project managers are misbehaving because their backs are against the wall. However, when the overall project is behind schedule and when assistant vice presidents begin to panic, you can be sure there is a lack of leadership.

Defining Leadership

Leadership is a word and concept that has been the basis of many books, capturing researchers' imaginations. The word is widely misunderstood,

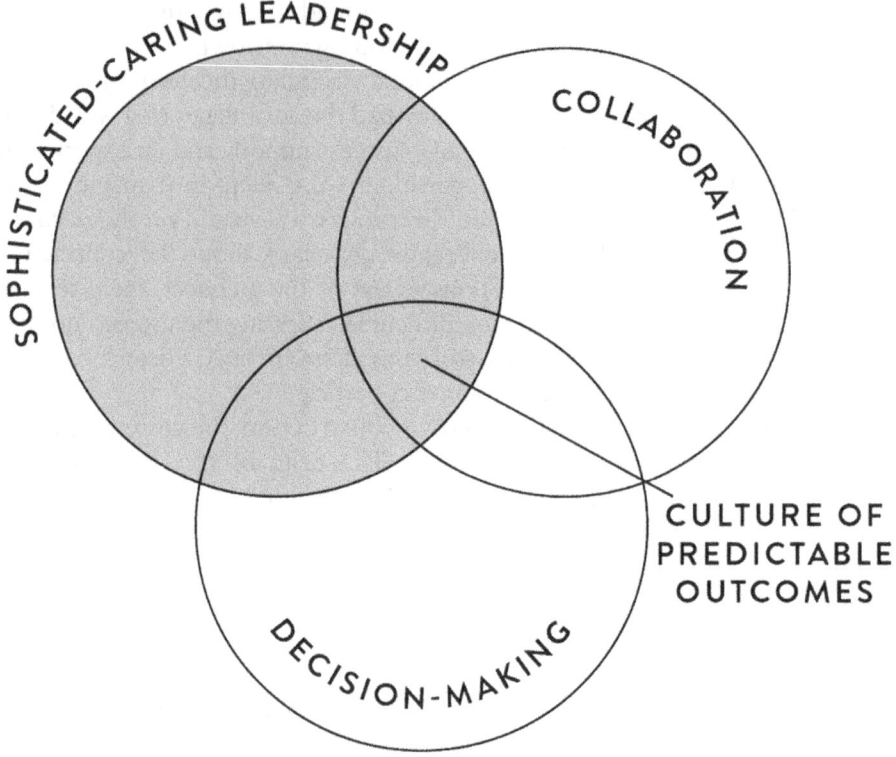

THE KEY INGREDIENTS
INCLUDE:

1. Demonstrating a Servant Leadership
   Mindset - Supporting Others to Success
2. Striving for Balance - Directing Clearly
   and Empowering Freely
3. Providing the Tools Required for the
   Journey
4. Motivating the Team
5. Coaching Team Members to Success or
   Coaching them Out the Door
6. Creating Psychological Safety

Figure 2.1 Key ingredients for sophisticated-caring leadership.

and yet, the characteristic is pursued by many. I have seen scores of bosses, unit heads, project managers, and team leaders that know nothing about leadership. Conversely, I have known administrative assistants, new hires, and students with a better grasp of leadership traits than those persons that outrank them in position or experience. Since there is so much confusion about the nature of good leadership, I will start with my definition.

Extraordinary leadership supports others to success by providing a clear path to travel, the tools required for the journey, and inspiration for the trip. Authentic leadership includes the courage to protect the team by removing bad or toxic players and creating an environment of psychological safety that serves the team's highest performance.

Sophisticated-caring leadership in the design and construction industries must have all these elements to be successful. The days of gotcha, bullying, adversarial leadership is over. This style of leadership has never served the industries well, and, in this age of disruption, it just looks ridiculous. The only leadership style that will be effective moving forward is sophisticated and caring, allowing transdisciplinary and transorganizational teams to thrive. No leader in the design and construction industries has a big enough hammer to force the industries to perform at their best. Only cooperative, inspired leadership that aligns goals among teams and organizations will accomplish that mission. That leadership style must focus on the following:

1 Demonstrating a Servant Leadership Mindset—Supporting Others to Success
2 Striving for Balance—Directing Clearly and Empowering Freely
3 Providing the Tools Required for the Journey
4 Motivating the Team
5 Coaching Team Members to Success or Coaching Them out the Door
6 Creating Psychological Safety

1 Demonstrating a Servant Leadership Mindset—Supporting Others to Success

John L. Hennessy, former president of Stanford University and chairman of Alphabet wrote, "For many people I know in positions of power and authority the hardest thing to learn—and some never do learn it—is that Leadership is service."[2] I agree with these words and have stated many times, sometimes to eye rolls, that supporting others to success is what gets me up in the morning. It doesn't matter whom I am helping, a student, a faculty member, an architect, an engineer, a contractor, or a family member. Supporting others and watching them succeed makes me happy. It's been a lucky dynamic for me because, as it works out, this characteristic also makes my organizations successful. I don't know precisely from where this personal sense of supportive leadership, often called servant leadership, came. Many of the role models I saw in the design and construction industries early in my career were traditional top-down, "just because I said so— read my mind," kinds of leaders. However, I did have a few mentors who demonstrated that supportive leadership delivered satisfying, meaningful, and beneficial results.

When I worked for the architectural firm Spillis, Candela, and Partners (now part of AECOM) in the 1980s and early 1990s, I managed a large regional headquarters project for USAA in Tampa, Florida. Ron Roeder was the project manager for USAA. Sam Ellison led the construction team for a company now called BECK, and Bob Crittenden led the efforts of Project Control, providing project management services. This project was my first experience with Total Quality Management (TQM) and Partnering, tools that are no longer fashionable but were successful in creating a team-focused experience. Most importantly, most of the time, the team leaders led in a supportive, friendly, and humble manner. They believed the success of the project was dependent on the success of the team. This lesson was vital for me as a young architect and project leader.

In 1995, I decided to leave private practice to work as the university architect for the University of Miami. I had no idea what transitioning into the owner's world would be like but was surprised and pleased to be mentored by two extraordinary professionals. The first was my direct supervisor, Joe Folino, who worked with me every day, not as a boss but as a partner. Also important, David Lieberman, senior vice president of Business and Finance, took every opportunity to coach me to success regarding communication style, strategic thinking, and preparation as I worked with internal clients, donors, and the board. Both of these men taught lessons of successful leadership.

The most important lesson I learned from these experiences was that my job as a leader on any project or within any organization was not to make people do what I said; rather, it was to provide my teams with whatever they needed to be successful. Over the years, I have found that responsibility includes providing information, decisions, coaching, guidance, motivation, partnership, confidence-building, emotional support, psychological safety, professional growth, rewards, discipline, and the truth. In other words, I am a service provider to my teams and every person on those teams. This type of leadership is hard work, but it is also rewarding and successful work.

In the design and construction industries as in higher education, asking leaders to be supportive, servant leaders is difficult because it is not part of our culture. As John L. Hennessy notes in *Leading Matters*,

> It is a difficult lesson because almost every other aspect of being a leader convinces leaders otherwise: leaders typically are paid more money than most of the people they lead, leaders hold authority over their teams, their decisions take priority, and their subordinates are ultimately dedicated to serving their leader (or more precisely, the institutions they lead). In the face of all that, it can be hard to remember that you, the leader, are serving them – they are doing the heavy lifting, while your job is to help make them as effective and productive as they can be.[3]

This point is essential. As a leader, you ask your team members, those on the front lines, to do the hard part, the heavy lifting. They need your help and support to be successful, and when they are successful, you are successful. Besides, your teams and team members have choices. They can follow you or not, by choosing other places to work, or simply by not working in the way you wish them to. According to Robert K. Greenleaf, a thought leader in servant leadership,

> A new moral principle is emerging, which holds that the only authority deserving one's allegiance is that which is freely and knowingly granted by the led to the leader in response to, and in proportion to, the clearly evident servant stature of the leader.[4]

For six years, I was privileged to be on the board of TDIndustries, a sizable mechanical subcontracting company in Texas and Arizona. TDIndustries is grounded in the servant leadership values of its founder, Jack Lowe Sr., and resiliency, regardless of economic cycles. Watching this company grow reinforced my views of valuing each team member, supporting team members to individual success, listening carefully to the needs of the team, and working as a servant leader even at the board level.

Navigating the future of the design and construction industries will be impossible without servant leadership. Changing generational values will ensure that autocratic leaders are left high and dry. Supportive leaders will be more agile and have a better chance of marshaling transdisciplinary, transorganizational teams.

2 Striving for Balance—Directing Clearly and Empowering Freely

In *The Owner's Dilemma*, I referred to the Three Cs, a tool I shared with my team when I first arrived at Rice University to run the major construction projects. My team needed guidance on how to manage their teams effectively, so I began with "Clarity, Cooperation, and Consistency," the Three Cs, as an easy-to-remember starting point. I asked the project managers to be *clear* with their teams regarding needs and expectations; to *cooperate* with their teams to supply information, promptly pay invoices, and provide decisions; and, finally, to be *consistent* in their demands as a leader. My observation had been that many project managers changed their minds regularly due to their own insecurities and lack of knowledge. The Three Cs pulled us through the early days of reorganization and of resetting our projects. As it turned out, providing Clarity, or a Clear Path to Success, to project the teams was a significant component to our success.[5]

Great leaders support their teams by defining success and a path to success. Often on project teams, that means determining the usual parameters: scope, budget, schedule, but it also includes a definition of how you are going to get there. This definition must outline the leader's expectations for process and communications. No team should have to read the mind of any leader. If a leader wants an update every Monday morning, the leader should state that requirement upfront. If a leader expects all team members to go to happy hour on Friday or to have invoices reconciled on the third Tuesday of the month, these requirements must be stated at the outset.

Most importantly, if you, as a leader, expect the team to engage in collaborative processes, you must be clear about your expectations. In *Teams Matter*, a 2016 study examining the benefits of specific practices of collaborative teams working on projects funded through the GSA American Recovery and Reinvestment Act (ARRA), Renée Cheng's recommendations included that teams set clear objectives and goals, and that they should define clear roles. Project teams provided with this leadership demonstrated better alignment, issue resolution, and decision-making. "When the GSA leadership set clear objectives and criteria for making decisions and explicit expectations for results, teams reported that they worked toward the same goals."[6] When this clarity existed, even when facing changing and complex environments, teams maintained their alignment.

However, in driving to this clarity, there is an enormous temptation to believe that the leader is simply the boss and what the leader says goes— just because they say so. Do not be deluded. Being a sophisticated leader means that you understand you are most often not the smartest person in the room on most subjects. Being an intelligent leader means that you will not have the best answer to every problem. Being a caring leader means that you are responsible for creating an environment where the right voices are heard, and the best solutions are found. Therefore, as the leader, you must find a balance between defining a clear path to success, encouraging curiosity and exploration, and enabling decision-making and action by others.

General Stanley McChrystal states this bluntly in his book *Team of Teams*,

> Although we intuitively know the world has changed, most leaders reflect a model and leader development process that are sorely out of date. We often demand unrealistic levels of knowledge in leaders and force them into ineffective attempts to micromanage. The temptation to lead as a chess master, controlling each move of the organization, must give way to an approach as a gardener, enabling rather than directing.[7]

This perspective is finally edging its way into our design and construction industries as evidenced by the book *Defining Contemporary Professionalism*, where London-based architect Alex Ely describes ideal leadership behaviors by also quoting General McChrystal,

Leaders are expected to empower their team, giving them a platform for developing ideas that work, as well as acting as an example to follow and offer direction—"The leader acts as an 'Eyes-On, Hand-Off' enabler who creates and maintains an ecosystem in which the organization operates."[8]

This balance between clarity and empowerment is critical in the twenty-first-century design and construction industries where the economy, technology, and even pandemics can throw disruption at us at any time.

Nowhere were the benefits of this dynamic, the balance between clear direction and enabling innovation, more apparent than during the Chilean mining disaster in 2010. In August of that year, thirty-three Chilean miners were trapped 2,300 feet, nearly half a mile, underground by two cave-ins. There was little hope that they would be found, let alone that they might survive. The world was transfixed while an international team, led by an extraordinary Chilean professional, gathered to save these men. Rescuing the miners was one of the most complex and impossible tasks imaginable, yet the team somehow pulled all thirty-three men alive from that mine sixty-nine days after the accident. Faaiza Rashid had been studying high-performing teams for several years at Harvard University alongside Amy Edmondson, a researcher in both the business school and social sciences. I knew both researchers through their work on studying construction teams and IPD processes. Faaiza had done some research on the Rice project teams and had written the chapter summary for *The Owner's Dilemma* on research methodology.

After the miners were pulled out of what might have been their tomb, Faaiza flew to Chile and interviewed the key members of the team. Her goal was to understand what might be learned from the Chilean events about leading highly complex dynamic teams in high-stress environments. Considering many design and construction projects meet this description—highly complex, dynamic, and high stress—perhaps we can find some applicable lessons. Your projects may not involve thirty-three men trapped deep underground; still, you may have millions of dollars and the safety of hundreds at stake each day while faced with the challenges of changing environments, changing information, and changing scope.

In Chile, Faaiza discovered a leadership dynamic that was impressive and very instructive. According to the article written by Faaiza along with Amy C. Edmondson and Herman B. Leonard and published in *Harvard Business Review* in 2013, the team had two distinct challenges. First, they had to find the miners. Then they had to determine how to rescue the trapped men if any were still alive. To accomplish each mission, they needed to encourage innovation but also to stay focused.

THE DUAL IMPERATIVES OF
HIGH-STAKES LEADERSHIP

		DIRECT	**ENABLE**
In time-pressured, make-or-break situations, leaders need to take a two-pronged approach, giving their teams clear direction and, at the same time, enabling rapid innovation. To implement the approach, they must perform three tasks, which encompass both kinds of activities.	**ENVISION**	Realistically assess the current situation and how it's likely to evolve	Articulate the possibility of hope against all odds
	ENROLL	Set boundaries for who will be on the team and motivate them to address the problem	Reach out to collaborate with diverse experts
	ENGAGE	Lead disciplined, coordinated execution	Invite innovation through experimentation and learning

Figure 2.2 From 'Leadership Lessons from the Chilean Mine Rescue'. Source: Faaiza Rashid, Amy C. Edmondson, Herman B. Leonard, "Leadership Lessons from the Chilean Mine Rescue," *Harvard Business Review*, July–August 2013, 10.

The authors stated,

> In such emotionally charged circumstances, most leaders feel torn. They worry: Should they be directive, taking charge and closely monitoring people? Or should they be empowering, inviting innovation and letting many experiments bloom? Our research suggests that the answer should be yes—to both. The choice presents a false dichotomy [...] To meet these conflicting demands, leaders must alternate between directing action and enabling innovation. At times, they must be decisive, give instructions, and periodically close down discussions so that the team can get things done. At other times, they must create space for new ideas, encourage dissent, ask questions, and promote experimentation. Leaders that lean too much toward either relentless commands or unchecked ideation do so at their peril (see Figure 2.2).[9]

Provide that clear path to success, but find the balance that allows teams to do their job, innovate, problem solve, perform at their best.

3 Providing the Tools Required for the Journey

Everyone that knew my father-in-law called him Rock, even his kids. A former colonel in the Marine Corps, Rock flew in three wars, starting with Corsairs in World War II. No one messed with Rock even though he had the kindest heart in the world. After the military, Rock retired to a beautiful farm in Colorado, again, long before it became the thing to do, and brought his seven children with him. He taught each of his children many lessons. One, in particular, stuck with me, "You need the right tool for the right job." That lesson applies equally to pitting cherries from the farm's orchard with an ancient but lovingly refurbished cherry pitter or managing a project with technology tools that support collaboration. Empower teams with the right tools.

Years ago, at Rice University, a board of trustees member, Ber Pieper, former president of Brown & Root, challenged me to create a dashboard on project status for the board's Buildings and Grounds Committee (B&G). I thought it was a great idea, but only if the task did not add additional stress and work for our project managers. Communicating with the B&G was a high priority, but creating multiple reporting processes was not a good idea. Multiple reporting processes created the opportunity for translation errors as well as redundant work. At the time, there was no software available that rolled up the project manager budgeting information to different levels of review so all projects could be reviewed in a consolidated yet understandable manner. So, we designed our own. The project managers had a tool that let them manage and report to the next level, and that same report could be rolled into the dashboard for the board of trustees. Trustees were able to access the reports with their own password. Each member of the team had the tool they needed to be successful. A decade later, these tools are available from national vendors, some modeled on the description in *The Owner's Dilemma*. The key is to find, use, and develop the tools your team needs.

As I stated earlier, great sophisticated leaders in the design and construction industries support others to success. In many ways, as the leader, you are the service provider for your teams working on the front lines of projects. Part of providing that service is providing the tools that the team needs to be successful. To do this, you must be curious. I have seen two kinds of leaders in the design and construction industries that fail when it comes to tools. The first kind of leader resists new tools because they constitute change. They blame the risk managers, and they blame the cost. These leaders don't have the time and patience to adopt new tools and processes. They have no patience to learn and no patience for change management.

Another type of failed leader will invest in every new technology tool they see or every new management process about which they read. This "oooh shiny" mentality is expensive and often unhelpful. These leaders load down their teams with tools misaligned with the needs of the project or the organization. These teams are overwhelmed with change and never leverage their tools fully or effectively.

Sophisticated-caring leaders listen to their teams and critically evaluate the needs of the projects and the individuals. These leaders collaboratively determine the tools required to serve these needs and constraints, including the budget, and they find ways of leveraging the full benefit of tool investments. Sophisticated leaders plan the rollout of new tools and processes, communicating broadly and proactively how the rollout will occur. Sophisticated-caring leaders invest in training and support to ensure the effectiveness of a tool or process rollout, and develop an evaluation process to make sure the tools are as useful as expected. Make sure your team has the right tools and processes to do the right job. Only leaders can make that happen.

4 Motivating the Team

Don Shula, former football coach of the Baltimore Colts and the Miami Dolphins, passed away on May 5, 2020. This news rocked my world. I moved to Baltimore from Long Island, New York, in 1969 just after the infamous Super Bowl III loss by Don Shula's Colts to the New York Jets led by quarterback Joe Namath. In subsequent years, I watched as Shula left Baltimore, followed by the retirement of Johnny Unitas and, ultimately, in 1983, the relocation of the Colts to Indianapolis. A short four years after his loss against the Jets, Shula led his Miami Dolphins to a perfect season and a win in Super Bowl VII. Perhaps the greatest coach in football history, he was, in part, the winningest because he was tough to play for and, in part, because he knew how to motivate his teams.

The day Shula died, Larry Csonka, Miami Dolphins Hall of Fame fullback, remembered,

> He changed everything when he got to Miami. He simply would not accept losing. To me, the most memorable moment with him wasn't any of our wins, but after a loss in Super Bowl VI. Right after the game, when it was only the players and coaches in the locker room, he said, 'I want every one of you to remember how you feel right now, so that you never feel this way again. I want to go into next year with the goal of doing everything we can to win each of the games we get ready to play.'[10]

The University of Arizona has a motivational football story of its own predating the famous story about Notre Dame legend Knute Rockne and the Gipper. In 1926, John Bryd "Button" Salmon was the football quarterback, baseball catcher, student body president, and the quintessential "big man on campus." Critically injured in a car accident during an excursion to Phoenix with some friends, the twenty-two-year-old Salmon exhorted his coach, J.F. "Pop" McKale, "Tell them [...] tell the team to bear down." The coach used the story and the phrase "Bear Down" to motivate his team, but this message also motivated the heartbroken student body. The phrase "Bear Down" lives on in fight songs, cheers, and even basic campus greetings.[11]

Coaches of sports teams understand how important it is to motivate their teams, to make sure they are moving with enthusiasm to a common goal. These coaches invest in motivation. Overcoming barriers to achieve any goal is hard work, and a team, along with the individuals on that team, must be motivated to that purpose. When a book opens with a compelling first few pages, we are motivated to keep reading. When our clothes fit too tightly or the doctor talks to us sternly, we are motivated to lose weight. When we sign up for the Boston Marathon, we are motivated to start training. When a murder is committed, detectives look for motive. Motivation drives us, compels us to take action, good or bad. Why is it then that so many leaders ignore this vital element? Why do so many leaders forget to consider motivation as a critical responsibility of sophisticated-caring leadership?

And why do leaders not understand the nuances of motivation? There is no one-size-fits-all approach when it comes to motivation. Sometimes a leader can create a **big vision** for a team, and sometimes a project comes with **intrinsic** motivation. Other times, leaders have to address motivating teams on **ordinary projects** and learn about **personal motivation**.

Big Vision Motivation

Why are we here? Why are we doing this? Why is it important? Why should I be willing to put up with the frustration, difficulties, and complexities of this project? If team members don't have a reason to do the hard stuff, they will back off, and they may shut down. Defining motivation is an absolute requirement for successful leadership of teams. Leaders that provide answers to these motivational questions gain teams that are personally committed to the project. Arthur E. Frazier, Spelman College architect, cited an example of a motivated and committed team in the book *Managing Design*,

> The Winship Cancer Institute at Emory went well from a collaboration perspective, and it had nothing to do with design or construction, it had to do with purpose. The Cancer Center representative stood up at the partnering meeting and said, 'Everybody in this room

has been touched by cancer.' It gave us a common purpose and theme for the entire project. We may not have been able to cure cancer personally, but we could do our best to give those who were trying the best building we could. It wasn't just a common purpose, it was a higher purpose. We repeated it when tradespeople came in, and at the topping out, we had cancer survivors sign the steel beam. It was inspiring. That was 1998, and I remember it like it was yesterday.[12]

Chuck Thomsen and Sid Sanders reflected on the importance of leaders sharing big inspiring visions in *Program Management 2.0*,

> Such leaders understand that an integrated team with fully aligned goals is a potent vision even if it can't be totally achieved. They understand the power of vision—that leaders have used idealism successfully to organize the behavior of people since the beginning of recorded history and that people will work dutifully for money but will work passionately for an idea.[13]

When team members are inspired and passionate, they are also more willing to take risks to be successful. Edmondson and Reynolds share in *Building the Future*, "Start with a Big Vision. Big Vision compels and motivates people- at different ages and from different backgrounds—to take personal and professional risks to pursue a dream."[14]

Intrinsic Motivation

There is no doubt, teams perform better when they have a reason—when they have a goal—when they are, together, motivated by a purpose. On March 27, 2020, only two weeks after the University of Arizona's COVID-19 isolation policies had begun to roll out, Kim Patten from the university's Research Development Office sent me a request for proposal (RFP) from the Regional Partnering Center (RCP). I had been working on developing collaborative built-environment research at the university, so Kim thought I might be interested in creating a team for this challenge. The RFP was for the development of mobile COVID-19 testing centers in a manner that would protect health care workers while at the same time testing large numbers of patients. Most of the critical criteria were still sketchy, yet the deadline for submission was just one week away. As I read the RFP, it occurred to me that the demands of this proposal were outrageous, yet they were entirely appropriate for the crisis at hand. I hoped that we might find a way to respond.

Kim and her team had created COVID-19 Slack sites and channels for internal communications about research related to addressing the virus. Before COVID-19, few on campus had used Slack. Now, dozens of sites and channels had erupted with activity. Many of the active discussions on these channels were related to manufacturing personal protective equipment (PPE) or medical-related research addressing virus treatment. Undaunted, I posted the RFP, asking for interested team members, and reached out to researchers that I had met across campus. Finally, I reached out to potential external partners that might be interested in the project. Angela Watson and Luke Voiland from Shepley Bulfinch responded immediately, as did Fred Knapp from Core Construction. Concurrently, my virtual on-campus appeals had resulted in a satisfying flurry of responses from potential team members from across campus, including the medical school. We had fourteen impressive experts and leaders from across campus anxious to get started when we met for the first time by Zoom at 4:00 p.m. on Monday, March 30, 2020.

I could tell from the start that the tone of this conversation was different from other teams I had been on, not just because of the Brady Bunch tiled world in which we engaged. The team was comprised of architects, mechanical engineers, and contractors, yes, but also clinicians, and process engineers, as well as researchers in community medicine, human wellness and performance, antimicrobial surfaces, complex systems, and biomedical equipment design.

Missing was the normal storming activity, jockeying for position, or individual call for attention that characterizes new teams. All participants seemed attentive and interested, even anxious, to find a way to contribute to this critical project. As I launched, I did not need to do much more than thank everyone for their interest. I did not need to provide an inspirational speech. The motivation was already present in that virtual room. As we collectively discussed the RFP, questioned various approaches, and, yes, challenged each other, there was no doubt that the team was aligned and wanted to make the team function. Two project leaders, Aletheia Ida and Altaf Engineer, emerged from the first two Zoom meetings. They grabbed the reins of the project. Again, they did not need to give motivational speeches. People were dying in other parts of the country. Five days later, five minutes before the deadline, our proposal was submitted, including concepts for two different approaches for mobile testing sites. We competed against four other teams, yet, two and a half days later, our proposal was selected by the RCP to move forward into full design. To those that understand the academic environment, this was nearly a miracle.

This story is an example of intrinsic motivation, the most powerful and compelling of team drivers. Leaders that recognize and leverage intrinsic motivation can move teams forward easily so long as the team actions align with the motivational goals. In the case above, the team found the challenge instantly persuasive. Every decision made aligned with the values of protecting patients and care workers while meeting the project's criteria. If cost or operational challenges had pushed the team to abandon the importance of protecting the patients or the caregivers, the team's intrinsic motivation could have been undermined, and the team's passion for the project quelled.

Thomsen and Sanders stated this dynamic concisely in *Program Management 2.0*, "A culture of motivation, shared goals, collaborative achievement and a sense of higher purpose will produce far better results than an authoritarian environment. People will work for money but they will fight and die for an idea."[15]

Motivating on the Ordinary Project

However, not every project has the benefit of a pandemic or a need for a cure to provide intrinsic or big vision motivation. It then falls to the team leaders to finds ways of developing engagement and motivation. Even on bank buildings, renovation projects, theaters, and retail facilities, it is necessary to conceive big goals that will help the team stay interested and excited about their work. Sid Sanders, senior vice president for construction, facilities, design, and real estate management for Houston Methodist, was asked by how he motivates his teams to innovate and take risks in such a large organization. Sanders responded,

> Despite our size, Methodist tends to be a pretty innovative organization itself. Whether you're a custodian or a senior executive, you're required to go through a two-day onboarding process to learn about our culture. The acronym ICARE is the core of our culture: Integrity, Compassion, Accountability, Respect. By the time someone completes this training, they understand that this is a unique place.

It sets the stage for saying, 'We want excellence in all we do and to create the best environments we can.'[16]

Consistent messaging and training is only part of creating motivation on projects some might think of as ordinary. Leaders must learn how to turn ordinary goals into extraordinary motivation. Thomsen and Sanders noted in *Program Management 2.0*,

> Each organization we talked to has different goals that influence outcomes. The classic triumvirate of cost, time and quality is usually there, but the emphasis—and the very definition of the terms—changes

with the organization's mission. Cost might mean first cost, return on investment or life-cycle cost. Time might mean as soon as possible. Or it can mean a date certain — like getting a school open for fall classes or a store open for the Christmas shopping season."[17]

This viewpoint is just a starting point for a sophisticated-caring leader. A school design might have in its mission statement that it is informed by research to improve teaching spaces at an affordable cost. An infrastructure project might set construction schedule goals that minimize user disruption. A classroom building may challenge its team to create the lowest long-term capital reinvestment by being flexible and easy to maintain. These goals become exciting and intriguing challenges to project teams and can become great motivators if supported by leaders.

Most importantly, leaders must make sure the goals motivating the team are clearly stated and aligned. Clear communication was a subject area studied by Renée Cheng and her team at the University of Minnesota. They investigated collaboration and leadership on GSA American Recovery and Reinvestment Act (ARRA) projects from 2004 to 2015. Goal alignment was demonstrated as a winning leadership strategy. On one of the study's most successful projects, "GSA leadership clearly articulated goals and ensured that the core-team members were all on the same page. Alignment among the members ensured clear and consistent messages on the ARRA goals, and project goals were regularly repeated during larger-team meetings."[18]

Sophisticated-caring leaders motivate their teams in a clear and compelling manner even when the projects seem mundane by defining challenging goals that will enhance the project and engage the team.

Personal Motivation

Leaders ignore the personal motivational needs of their team members at their peril. Each team member is an individual with unique and personal needs, competencies, and aspirations. Many leaders do not understand that monetary rewards are not sufficient for personal motivation and may even be counterproductive. One research study documented by Uri Alon provided the following results:

> People are given interesting mechanical puzzles to solve. Group A is given a dollar for solving each puzzle; Group B is not. After 30 min, the researchers tell the groups that the experiment is done. It was found that Group A puts the puzzles down, whereas Group B keeps playing

with them on their own time. The surprise was that money and other rewards in these types of tasks apparently act to reduce motivation. What makes people motivated?[19]

According to Alon, members of teams, in this case, research teams, are personally motivated by three specific characteristics, including

- confidence in their competence to complete a task,
- autonomy to self-direct and problem solve during a task, and
- social reinforcement from someone on the team that cares about the individual personally and their success.

An example of the first personal motivation characteristic comes from *The Girls of Summer*.

> When [Briana] Scurry was assigned to start her first match in goal, at a tournament in Portugal in 1994, she said that [Carla] Overbeck [the team captain] approached her and said, "You deserve to be here." "That really helped," Scurry said. "I wasn't sure I was ready yet. I went from number five on the list to number one in three months [...] It was hard for me to judge me myself, but she obviously had faith in me. I figured that if she did, I must be good enough.[20]

It is also important that the leader appropriately challenge team members, giving each work to stretch their abilities and build confidence but not overwhelm or discourage them. However, to some leaders I also warn, challenge young professionals bravely; they are hungry for opportunities to prove they can learn, fail, and learn from failure—just as you did when you were a young professional.

The ability to fail speaks to personal motivation number two. Stand back and let team members learn to solve problems. When a team member asks you for assistance, don't jump ahead to solve the problem yourself. Provide some guidance but encourage the team member to find their solutions on their own. Help them grow and gain the satisfaction that goes along with this kind of accomplishment.

The third characteristic of personal motivation is most appropriately illustrated by the great Don Shula when he provided opportunities for team members to recognize the accomplishments of their peers.

> Years ago, I started a meeting pattern to help recognize the less-publicized players. The day after a ball game, our team and coaches review our

performance. I will make some opening remarks to the squad, critiquing what happened during the game—good, bad, or otherwise. Then the entire squad views the game films that focus on our special teams. We use this time to create opportunities for players to appreciate each other's efforts. It makes special team players feel important when a star like Dan Marino says, "Hey, that was a great hit!"[21]

Sophisticated-caring leaders find the time to support and reinforce the efforts of individual team members and to support the aspirations of those individuals. These leaders never forget personal motivation is as important as the Big Vision.

5 Coaching Team Members to Success or Coaching Them Out the Door

One of the most difficult yet most pressing challenges of any sophisticated-caring leader is the process of hiring, coaching, and firing members of their teams. Before we dive into the details of this process, it is essential to reflect upon the nature of contemporary project teams in the design and construction industry. Because of the fragmented and fluid status of the industry, every project calls for a new set of team members. In addition, members flow from team to team, often changing mid-project. This dynamic condition means the benefits of standard, stable teams, similar to sports teams, are hard to leverage. Frankly, there are few stable, consistent teams in the design and construction industries with the ability to take advantage of long-term relationships, deep knowledge of working styles, and mutual trust. Amy C. Edmondson and Jean-Francois Harvey note that there is a difference between standard stable teams and contemporary "teaming," which is a process that occurs across boundaries in more fluid environments. In their book *Extreme Teaming*, the authors consider research "to understand what leaders can do to support teamwork in shifting configurations and contexts, including teamwork that brings people from different organizations together on a novel project."[22]

In this teaming environment, it becomes even more critical to make sure team members are aligned, equipped to do their work, and supportive of other team members. Team members must also share accountability and the values of the team. If any team member does not demonstrate these characteristics, it is the leader's responsibility to address problems quickly and effectively. Be assured, other team members are watching. If you want your team to accept the values and follow the processes you have defined together, it is critical to address individual team behaviors and competency issues as they arise.

Hiring Carefully

Leaders must take the time to carefully hire or select team members. The need to hire carefully is not new news. Back in 2001, when Jim Collins published *Good to Great*, we all became familiar with the concept of "first getting the right people on the bus (and the wrong people off the bus)." Few people remember that one of Collins's main points was "'who' questions come before 'what' decisions—before vision, before strategy, before organization structure, before tactics."[23] In the design and construction industries, the "who" generally comes way down the list, as if anyone can do what we do. The ironic thing is that we all know that who we work with makes all the difference in our success, yet we do not change our behavior, our processes, or our industry to make sure we work with team members aligned with our goals.

Even within conventional design and construction delivery processes, projects fail because of the perceived need to expedite selection processes or to settle for the best available team member even though the team member is not a good fit. Select team members carefully regardless of the circumstances because whatever time or money you believe you are saving will be lost due to the behavior or competency challenges that toxic team members present. According to Sid Sanders of Methodist Hospital, selecting team members well can almost be a superpower,

> I would say my leadership philosophy boils down to finding the right people, getting them into the right place with the right resources, and then getting out of their way [...] So, to be successful, I look for people I can rely on to oversee the design and construction process, where I can trust them to take millions of dollars and turn them into exceptional facilities, that will add value and long-term vitality to the institution.[24]

The lesson of carefully selecting team members is demonstrated over and over again in industry after industry. For example, at the Mayo Clinic

> an investment in employees is an investment in success for both the individual and the organization. In service companies, the service is a performance, and the employees are the performers [...] hiring the right people is the first rule; supporting and rewarding them is its corollary. Some basic criteria for evaluating potential employees include: Personal values complement the company's foundational values. Attitude is amenable—has willingness to fit into the organization as it is but has the courage to challenge status quo if necessary. Is talented in professional skills and in teamwork. Has potential to grow

and develop expertise beyond present level. Is interested in a career, not just a job. Is loyal to employer. Through deliberate hiring based on a rigorous screening process, interviewers find excellent people who will bring the production to life.[25]

These have been Mayo Clinic standards for many years, even prior to Mayo becoming a household name in health care.

Defining the characteristics of a great team member is critical and unique to every organization and team. Patrick M. Lencioni took on this subject in his book *The Ideal Team Player: How to Recognize and Cultivate the Three Essential Virtues*, where a construction company in trouble tries to figure out how to expand its teams effectively. Some leaders' first inclination might be to find the smartest person possible; however, Lencioni noted, "Keep in mind that being smart doesn't necessarily imply good intentions. Smart people can use their talents for good or ill purposes. In fact, some of the most dangerous people in history have been noted for being interpersonally smart."[26] Therefore, the book's parable advises,

First, we go figure out how to recognize a real team player, the kind of person who can easily build trust, engage in healthy conflict, make real commitments, hold people accountable, and focus on the team's results. Then, we stop hiring people who can't.[27]

Coaching Consistently

If you already have a team member misbehaving on your team through nonperformance or in some manner toxic to team culture, act quickly to coach that person to success. The best approach and investment will be to set up a coaching process for every member of your team from the start of the project, so they are all getting consistent feedback of all kinds. Keep these ordinary sessions focused on improvement. If an individual needs specific correction, it will be easy to make that correction within the regular coaching sessions and to make the consequences of the behavior clear. Always place a time limit on the expected correction by the team member. Act on that time limit. If the problem is competency, develop a plan for professional development, if there is time, or find a way to provide that individual with additional support needed, if the individual is on a growing trajectory.

The most crucial aspect of coaching is just to do it, even when it is difficult or uncomfortable. In one of my favorite books about leadership, *Dare to Lead*, Brené Brown pointedly observes,

We avoid tough conversations, including giving honest, productive feedback. Some leaders attributed this to a lack of courage, others to a lack of skills, and, shockingly, more than half talked about a cultural norm of "nice and polite" that's leveraged as an excuse to avoid tough conversations. Whatever the reason, [in Brown's research] there was saturation across the data that the consequence is a lack of clarity, diminishing trust and engagement, and an increase in problematic behavior, including passive-aggressive behavior, talking behind people's backs, pervasive back-channel communication (or "the meeting after the meeting"), gossip, and the "dirty yes" (when I say yes to your face and then no behind your back).[28]

If it is clear the behavior of an individual team member is not responding to purposeful coaching, or if the competence of an individual or firm is clearly below the standard required for the project, the leader must act swiftly to separate that individual or firm from the team. In short, the leader must be willing to fire team members that are not adding value. Even though these steps are taken only after appropriate coaching, firing team members is an unhappy and messy business. This action is part of being a sophisticated-caring leader in the design and construction industries. As Caroline Buckingham and Karen Mosley stated in *Defining Contemporary Professionalism*, "A leader must be able to make tough decisions, stand tall and be accountable—qualities and skills that individuals are not generally born with, but which can be learned."[29] Perhaps Patrick Lencioni said it more precisely, "we help the people who are acting like jackasses change their ways or move on to different companies."[30] Sophisticated-caring leaders care for their teams by hiring wisely, coaching thoughtfully, and firing when needed.

6 Creating Psychological Safety

Margaret Heffernan wrote in TED.com, "Grumpy orchestras tend to play better than cheerful ones; they're focused on performing better, and happiness is the output, not the input, of their work together."[31] To me, this comment hits exactly the right note. I can recognize the sound of a high-functioning team in the design and construction industries within a few moments of walking in the room. The first thing I notice is the team members couldn't care less about me, regardless of my role. The next thing I notice is the chatter, the informal, easy manner team members, irrespective of status or discipline, engage each other. There is usually humor and good-natured jibes, but mostly there is hard work and clear speaking. It's often the tempo of the conversation that is the best giveaway. High-performing team members listen when others speak, stop when tempted to interrupt, but there is no hesitancy to contribute or even to challenge when appropriate. In my experience, high-performing teams have high energy and little patience for distractions or nonsense. However, it is always safe for any member of the team to offer a crazy idea to solve a problem or to innovate.

We Need the Truth

High-performing teams in many different industries have these characteristics. It is no accident these characteristics are found in environments defined as providing psychological safety. In *The Fearless Organization*, Amy C. Edmondson stated,

> Psychological safety describes a belief that neither the formal nor informal consequences of interpersonal risks, like asking for help or admitting a failure, will be punitive. In psychologically safe environments, people believe that if they make a mistake or ask for help, others will not react badly. Instead, candor is both allowed and expected. Psychological safety exists when people feel their workplace is an environment where they can speak up, offer ideas, and ask questions without fear of being punished or embarrassed.[32]

Unfortunately, these characteristics are not typical in all teams. Many teams are hamstrung by fear, fear of being misunderstood, fear of taking risks, and fear of losing face. As a result, many project teams do not benefit from team members' knowledge and experience. Edmondson wisely observed,

> Knowledgeable, skilled, well-meaning people cannot always contribute what they know at that critical moment on the job when it is needed. Sometimes this is because they fail to recognize the need for their knowledge. More often, it's because they're reluctant to stand out, be wrong, or offend the boss. For knowledge work to flourish, the workplace must be one where people feel able to share their knowledge![33]

Brené Brown agrees,

> Not enough people are taking smart risks or creating and sharing bold ideas to meet changing demands and the insatiable need for

innovation. When people are afraid of being put down or ridiculed for trying something and failing, or even for putting forward a radical new idea, the best you can expect is status quo and groupthink.[34]

Design and construction projects are enormously complex and are becoming more so every day. The challenges facing the design and construction industries make it clear that multidisciplinary teams, not individuals, are needed to deliver these projects. Walter Isaacson advocated for building safe collaborative environments in the foreword to Leading Matters, observing, "The four seminal innovations of the digital age—the transistor, the computer, the microchip, and the packet-switched network—were all developed by collaborative teams rather than singular inventors."[35] So, creating environments where every team member feels supported to add their highest value is absolutely critical for the design and construction industries. The responsibility to create psychologically safe environments falls not only to organizational leaders but to team leaders as well. Edmondson states straightforwardly, "psychological safety is very much shaped by local leaders."[36]

In their own firm, Caroline Buckingham and Karen Mosley learned that

being empathetic—listening and understanding others—and having the ability to manage emotions were key skills that we began to bring to the fore. Simultaneously, having a clear and shared vision was critical to success. We worked hard to challenge previously existing internal barriers and silos, enabling teams to work in a more collaborative, transparent and open environment.[37]

The Grumpy Truth

When I have used the term "psychological safety" in the past, I have seen heads in the design and construction industries shake and eyes roll. Some people misunderstand this phrase, believing that providing psychological safety means that you are compromising excellence to create a nice, polite environment. Nothing could be further from the design and construction industry DNA, and nothing could be further from the truth. Psychological safety is all about creating environments where excellence and innovation are most likely to be created. Not surprisingly, exchanges may even be a bit grumpy. To achieve excellence, team members must be able to tell each other the truth, especially when they are making mistakes. Psychological safety is not about being nice.[38]

According to Harvard researcher Gary P. Pisano, "Psychological safety is an organizational climate in which individuals feel they can speak truthfully and openly about problems without fear of reprisal."[39] This type of culture is extraordinarily important for any team in complex environments because stuff goes screwy so quickly when team members do not tell each other the truth. "If people are afraid to criticize, openly challenge superiors' views, debate the ideas of others, and raise counter perspectives, innovation can be crushed."[40] Telling the truth in the moment, when it is most impactful, can make a team more successful. Again, Pisano observes,

> When it comes to innovation, the candid organization will outperform the nice one every time. The latter confuses politeness and niceness with respect. There is nothing inconsistent about being frank and respectful. I would argue that providing and accepting frank criticism is one of the hallmarks of respect. Accepting a devastating critique of your idea is possible only if you respect the opinion of the person providing that feedback.[41]

In the design and construction industries where we are integrating multiple disciplines, advancing technologies, and dealing with ever-changing teams while designing and building highly complex one-of-a-kind structures, it is essential that team members feel safe to tell the truth and share ideas. Creating these environments is not easy. Brené Brown urges,

> we have to cultivate a culture in which brave work, tough conversations, and whole hearts are the expectation, and armor is not necessary or rewarded. If we want people to fully show up, to bring their whole selves including their unarmored, whole hearts—so that we can innovate, solve problems, and serve people—we have to be vigilant about creating a culture in which people feel safe, seen, heard, and respected. Daring leaders must care for and be connected to the people they lead.[42]

Creating this kind of culture is the antithesis of a typical hierarchal design and construction project. Leaders will have to be purposeful and committed about supporting a culture that will encourage truth-telling and encourage risk-taking.

Do not be fooled into believing that this is a fantasyland culture or that leaders do not have the ability to manage underperformers. As Amy Edmondson states, "Psychological safety is not immunity from consequences, nor is it a state of high self-regard. In psychologically safe workplaces, people know they might fail, they might receive performance feedback that says they're not meeting expectations, and they might lose their jobs due to changes in the industry environment or even to a lack of competence in their role [...] But in a psychologically safe workplace, people are not hindered by interpersonal fear (See Figure 2.3)."[43] Gary Pisano also reminds us that these cultures create places where there is pushback,

> If it is safe for me to criticize your ideas, it must also be safe for you to criticize mine—whether you're higher or lower in the organization than I am. Unvarnished candor is critical to innovation because it is the means by which ideas evolve and improve.[44]

No doubt, creating environments of high psychological safety is a balancing act for leaders. Being candid serves the goals of excellence and innovation but can easily tip to disrespect and hurt. As Pisano advocates,

> Senior leaders need to set the tone through their own behavior. They must be willing (and able) to constructively critique others' ideas without being abrasive. One way to encourage this type of culture is for them to demand criticism of their own ideas and proposals.[45]

The outcomes are well worth the effort. Not only does an environment of psychological safety encourage excellence and innovation, but it increases learning. When team members are fearful and distracted by behaviors that make them protect themselves, they are not open to new experiences or new ideas. Increased learning feeds further innovation, excellence, and

HOW PSYCHOLOGICAL SAFETY RELATES TO PERFORMANCE STANDARDS

	LOW STANDARDS	HIGH STANDARDS
HIGH PSYCHOLOGICAL SAFETY	Comfort Zone	Learning & High Performance Zone
LOW PSYCHOLOGICAL SAFETY	Apathy Zone	Anxiety Zone

Figure 2.3 From the fearless organization: Creating psychological safety in the workplace for learning, innovation, and growth.
Source: Amy C. Edmondson, *The Fearless Organization: Creating Psychological Safety in the Workplace for Learning, Innovation, and Growth* (Hoboken, NJ: Wiley, 2019), 775.

confidence. Psychological safety creates an environment that challenges every member of the team to be brave and take risks. Expectations are high, and as Brené Brown protests, "If you are not in the arena getting your ass kicked on occasion, I'm not interested in or open to your feedback."[46]

Summary

Sophisticated-Caring Leadership is defined very differently from the traditional leadership model of the design and construction industries. A leader supporting a Culture of Predictable Outcomes demonstrates a Servant Leadership mindset by supporting others to success, strives for balance by directing clearly and empowering freely, provides teams the tools required to do a great job, motivates the team, coaches team members to success or coaches them out the door, and creates an environment of psychological safety.

Notes

1 Jere Longman, *The Girls of Summer: The U.S. Women's Soccer Team and How It Changed the World* (Pymble, Australia: HarperCollins e-books, 2000), 2325–2360.
2 John L. Hennessy, *Leading Matters: Lessons from My Journal* (Stanford, CA: Stanford Business Books, 2018), 35.
3 Ibid.
4 Robert K. Greenleaf, *Servant Leadership: A Journey into the Nature of Legitimate Power and Greatness* (New York/Mahwah, NJ: Paulist Press, 1977).
5 Barbara White Bryson and Canan Yetmen, *The Owner's Dilemma: Driving Success and Innovation in the Design and Construction Industry* (Atlanta: Ostberg Library of Design Management, Greenway Communications, 2010), 63–64.
6 Renée Cheng, "Teams Matter: Lessons from ARRA, GSA Region 5 and the American Recovery and Reinvestment Act," School of Architecture, University of Minnesota, GSA Region 5, 4240 Architecture, May 2015, 21.
7 Stanley McChrystal, Tantum Collins, and David Silverman, *Team of Teams* (New York: Penguin Publishing Group, 2015), 232.
8 Alan Jones and Rob Hyde, eds., *Defining Contemporary Professionalism: For Architects in Practice and Education* (London: RIBA, 2019), 90.
9 Faaiza Rashid, Amy C. Edmondson, and Herman B. Leonard, "Leadership Lessons from the Chilean Mine Rescue," *Harvard Business Review*, July–August 2013, 113–119
10 "Dolphins, NFL Family Pay Tribute to Don Shula," Miami Dolphins, www.miamidolphins.com/news/don-shula-remembered-by-players-nfl-executives.
11 "The Story of Bear Down," The Arizona Wildcats, May 12, 2020, https://arizonawildcats.com/sports/2013/4/18/208236191.aspx.

12 Michael Alan LeFevre, *Managing Design: Conversations, Project Controls, and Best Practices for Design and Construction Projects* (Hoboken, NJ: Wiley, 2019), 31.

13 Chuck Thomsen and Sid Sanders, *Program Management 2.0: Concepts and Strategies for Managing Building Programs* (rev. ed.) (McLean, VA: CMAA, 2011), 35–36.

14 Amy C. Edmondson and Susan Salter Reynolds, *Building the Future: Big Teaming for Audacious Innovation*, (Oakland, CA: BK/Berrett-Koehler, 2016), 185.

15 Chuck Thomsen and Sid Sanders, *Program Management 2.0: Concepts and Strategies for Managing Building Programs* (rev. ed.) (McLean, VA: CMAA, 2011), 558.

16 Hannah Chenoweth, "Leadership Series: Sid Sanders, Houston Methodist. We spoke to Sanders about building strong teams and got takeaways from his book, 'Program Management 2.0,'" *Health Spaces*, July 25, 2019, https://info.healthspacesevent.com/blog/leadership-series-sid-sanders-houston-methodist.

17 Chuck Thomsen and Sid Sanders, *Program Management 2.0: Concepts and Strategies for Managing Building Programs* (rev. ed.) (McLean, VA: CMAA, 2011), 76–77.

18 Renée Cheng, "Teams Matter: Lessons from ARRA, GSA Region 5 and the American Recovery and Reinvestment Act," School of Architecture, University of Minnesota, GSA Region 5, 4240 Architecture, May 2015, 29.

19 Uri Alon, "How to Build a Motivated Research Group," *Molecular Cell* 37, no 2 (2010): 151–152.

20 Jere Longman, *The Girls of Summer: The U.S. Women's Soccer Team and How It Changed the World* (Pymble, Australia: HarperCollins e-books, 2000), 2360–2370.

21 "Don Shula's over-learning approach to coaching," Coach & A.D., March 28, 2018, https://coachad.com/articles/don-shulas-overlearning-approach-to-coaching/.

22 Amy C. Edmondson and Jean-Francois Harvey, *Extreme Teaming: Lessons in Complex, Cross Sector Leadership* (Binley, UK: Emerald Publishing, 2017, Kindle ed.),192.

23 Jim Collins, *Good to Great: Why Some Companies Make the Leap…and Others Don't* (New York: Harper Business, 2001), 63.

24 Hannah Chenoweth, "Leadership Series: Sid Sanders, Houston Methodist. We spoke to Sanders about building strong teams and got takeaways from his book, 'Program Management 2.0,'" *Health Spaces*, July 25, 2019, https://info.healthspacesevent.com/blog/leadership-series-sid-sanders-houston-methodist, 3–5.

25 Leonard L. Berry and Kent D. Seltman, *Management Lessons from the Mayo Clinic: Inside One of the Most Admired Service Organizations* (New York: McGraw Hill, 2008, Kindle ed.), 2889.

26 Patrick Lencioni, *The Ideal Team Player: How to Recognize and Cultivate the Three Essential Virtues*, a Leadership Fable (Hoboken, NJ: Jossey-Bass/Wiley, 2016), 160–161.

27 Ibid., 61–62.

28 Brené Brown, *Dare to Lead: Brave Work. Tough Conversations. Whole Hearts.* (New York: Penguin Random House, 2018), 7–9.

29 Alan Jones and Rob Hyde, eds., *Defining Contemporary Professionalism: For Architects in Practice and Education* (London: RIBA, 2019), 47.

30 Patrick Lencioni, *The Ideal Team Player: How to Recognize and Cultivate the Three Essential Virtues, A Leadership Fable* (Hoboken, NJ: Jossey-Bass/Wiley, 2016), 62.

31 Margaret Heffernan, "The Secret Ingredient That Makes Some Teams Better Than Others," Ideas.ted.com, 2015, https://ideas.ted.com/the-secret-ingredient-that-makes-some-teams-better-than-others/.

32 Amy C. Edmondson, *The Fearless Organization: Creating Psychological Safety in the Workplace for Learning, Innovation, and Growth* (Hoboken, NJ: Wiley, 2019), 720–727.

33 Ibid., 332–334.

34 Brené Brown, *Dare to Lead: Brave Work. Tough Conversations. Whole Hearts.* (New York: Penguin Random House, 2018), 7–9.

35 John L. Hennessy, *Leading Matters: Lessons from My Journal* (Stanford, CA: Stanford Business Books, 2018), xi.

36 Amy C. Edmondson, *The Fearless Organization: Creating Psychological Safety in the Workplace for Learning, Innovation, and Growth* (Hoboken, NJ: Wiley, 2019), 661.

37 Alan Jones and Rob Hyde, eds., Defining Contemporary Professionalism: For Architects in Practice and Education (London: RIBA, 2019), 45.

38 Margaret Heffernan, "The Secret Ingredient That Makes Some Teams Better Than Others," Ideas.ted.com, 2015, https://ideas.ted.com/the-secret-ingredient-that-makes-some-teams-better-than-others/.

39 Gary P. Pisano, "The Hard Truth about Innovative Cultures," *Harvard Business Review,* January–February 2019, https://hbr.org/2019/01/the-hard-truth-about-innovative-cultures?...4a50d3&hctky=11458115&hdpid=c8634 bfb-8e7d-4cb7-8c3a-2337818730b9, 9.

40 Ibid.

41 Ibid.

42 Brené Brown, *Dare to Lead: Brave Work. Tough Conversations. Whole Hearts.* (New York: Penguin Random House, 2018), 12.

43 Amy C. Edmondson, *The Fearless Organization: Creating Psychological Safety in the Workplace for Learning, Innovation, and Growth* (Hoboken, NJ: Wiley, 2019), 352–359.

44 Gary P. Pisano, "The Hard Truth about Innovative Cultures," *Harvard Business Review,* January–February 2019, https://hbr.org/2019/01/the-hard-truth-about-innovative-cultures?...4a50d3&hctky=11458115&hdpid=c8634 bfb-8e7d-4cb7-8c3a-2337818730b9, 9–10.

45 Ibid., 10–11.

46 Brené Brown, *Dare to Lead: Brave Work. Tough Conversations. Whole Hearts.* (New York: Penguin Random House, 2018), 20.

Chapter Supplement—2
When Leaders Want Answers

Note: Names of participants, firms, and institutions have been changed at the request of the participants.

MSVN's website declares, "We Build Anything." In truth, the company, originally founded by William Masterson Sr. in 1950, on April Fool's Day, to build homes and offices after World War II, today provides an extraordinary range of services from planning to facilities management. Now, seventy years later, the company of more than 5,000 employees, is a national multi-service powerhouse with offices in eleven cities, although construction is still a significant focus. The company's leadership team had taken on many challenges over the last decades, and few problems frustrated them as much as the subject of high-performing teams. When it came to comprehending why some teams succeeded and why some teams failed, MSVN's leadership was at a loss.

MSVN's leadership did understand they had some tremendously successful teams, but they also had a few unpredictably unsuccessful teams. Whatever the "secret sauce" of success was for building high-performing teams consistently, it did not seem to include simply assembling the best and brightest professionals on a single team. The leadership had tried that experiment more than a few times only to be deeply disappointed at the results. This challenge had been on the top of the leadership list for five years, but solving the problem of consistently creating high-performing teams would require a special task force with unique skills.

Six professionals across the company were tapped for this effort and challenged with helping the company understand better why teams succeed or fail, and what might be done to make more teams predictably successful. The task force was made up of a diverse group of employees from across the country. The members came from different offices and held various roles within the company. This newly formed team, this task force, was given only ten months to come up with answers.

Task Force

Nancy Rice reflected,

> I think when they [the MSVN leadership] put our team together. They saw different traits [after a personality assessment], and we all had different areas of responsibility. The leadership purposefully put the six of us together and gave us this squishy topic assuming we could begin to break it down, figure out what's at the root of it, and be able to articulate how you go about and provide recommendations.

The Teaming Task Force included the following members:

- Nancy Rice, director of business development in the Houston office
- Betty January, project manager for the Federal Group in Minneapolis
- Terrance Reins, project executive for the Sports Group, headquartered at the Kansas office
- Paul Dinsmore, senior superintendent in New York (now field operations manager in Seattle)
- Sean Miller, general manager of the Civil Group in Dallas
- Peter White, senior project manager for the Sports Group in Kansas, but worked all across the country

The task force met regularly. One time per month, the task force met in person for two days. The rest of the time, the team met virtually. They used "One Note" software to share research and prepare for meetings. Conference calls were held at least once a week when they discussed progress. The team used both quantitative and qualitative methods to do their work. MSVN already had four key metrics in place for assessing team performance during projects and at completion. Safety is the primary, most important, metric, followed by customer satisfaction (compared to MSVN peers), team experience, and, of course, margin performance compared to proposed or budgeted margin.

Team experience is defined as the experience of the individuals within the team during the project delivery. The task force used these existing metrics to identify high-performing teams and teams that were not performing as well. Then, they conducted interviews with as many teams as possible, looking for patterns.

According to Nancy Rice,

> We split into two teams of three for project site interviews, so one team of three [focused on] one part of the country and a list of

projects [including] both high-performing and low-performing teams. Then the other team of three visited the other [teams]. Although you could take the liberty to chase something down if something was interesting, we kept the questions consistent. So as we interviewed all of the team members on [different teams], we had some consistency, and then we were able to marry the results of the interviews against some of our key metrics the company uses.

The group quickly recognized that the challenge was different for stable business unit teams and project teams. Business teams and other administrative teams were together consistently over more extended periods, even years. They had plenty of time to get to know each other, work out processes, align goals, smooth communications, and learn how to collaborate. The 250 or more project teams that are created annually at MSVN have a much different challenge. These teams are required to start performing at a high level very soon after forming, so the task force decided to focus on project teams. Nancy Rice noted,

> Normally when you have a leadership team, they have time to start to build the trust, coalesce, and be able to transform to that high-performing side. However, we found from our historical data within our company that most of these project teams only operate together for an average of 249 days [...] So [our task force wanted to know] how do you quickly move into high-performing [mode], if your entire duration of time together is less than a year? [We asked ourselves] how do you really expedite the team's ability to get through the Forming, Storming and Norming phases [of team building] to get to Performing?

The task force also realized there were many reasons why teams succeeded and failed. Each project had unique conditions, weather, and partners. Sometimes failure seemed unavoidable when trade partners failed or went out of business, or weather caused significant delays or damage. The failure of some projects seemed unfathomable, where a great team failed for no tangible reason. Even so, after weeks of work by the task force, a few patterns began to emerge. For example, the task force was able to substantiate the MSVN leadership anecdotal belief that you can't simply build a high-performing team by putting together a group of high-performing individuals, a common practice for many organizations. Nancy Rice noted,

> With a big complex project, [we] throw all of our best people together. [We might] put them on this $1.5 billion project, and [assume] it is going to be a high-performing team because every individual on that

team is a high-performing individual. [The task force found] that's not the case.

We also found some of the biggest obstacles of low-performing teams were lack of clarity of roles and lack of alignment of roles and responsibilities. Leadership had certain expectations [but they weren't being articulated]. Nobody goes into their job thinking they are trying to fail or even on the path to fail, but the misalignment of clarity in roles and responsibilities provided a lot of obstacles for a number of team members that were in that low performing camp.

Another pattern emerged related to a surprising quirk in the data. The task force noted that some projects, early in the schedule (at 5 percent completion), might score high on safety, customer experience, and margin, but score low on team member experience. As they tracked these projects to completion, the task force noted that these teams would almost inevitably fail. As it turned out, the team member experience metric was an early indicator of potential team performance problems.

Psychological Safety

About the time the task force recognized the importance of the team member experience metric, they also became aware of a concept called psychological safety through a series of interviews with a project team at Google working on Aristotle 2.0 in 2017. Rice remembered the first conversations made psychological safety sound "really squishy," especially since there wasn't much published about the subject at the time. But after reviewing the outcomes from the Aristotle 2.0 teams and the work of Harvard researcher Amy Edmondson, the task force members thought they might be on to something. Based on the task force research, psychological safety created an environment where team members felt they could share ideas generously, open themselves to learning, and trust other team members.

Wondering how the creation of psychological safety might impact project teams in an accelerated teaming environment, the task force began to ask more questions. Rice noted,

> If you look at a company of 5000+ full time employees putting together all of these project teams, a project team may have anywhere from two people to 50 people, and you have to ask how does an average team of 10 or 12 suddenly build psychological safety, collectively, not just two people but all 10–12? And then what does that really mean?

Through the research and interview process, the task force members started pulling apart the concept of psychological safety into four different "buckets" including Learner Safety, Challenger Safety, Inclusion Safety, and Collaborator Safety.

- **Learner Safety** is the willingness to learn, share ideas, and look for opportunities to break through, ask questions, try new things.
- **Challenger Safety** is being able to challenge someone on your team regardless of your role, tenure, or hierarchy, and it is safe to challenge without retribution. Challenges will be listened to and entertained.
- **Inclusion Safety** is agnostic of title position, authority background, or experience, and means you will be included in the conversation and your voice is valued.
- **Collaborator Safety** relates to mutual access, to innovate together.

Nancy Rice observed that Inclusion Safety overcame titles, expertise, or even presumed areas of responsibility. For example,

> if your primary focus is estimating and the team is trying to figure out the best place for the crane pick, if you offered a suggestion or have a particular perspective, the team will not disregard your perspective because you're the estimator. You may not be the superintendent who has set a crane pick 50 times, but you will be heard even if you are the estimator or the intern.

Collaborator Safety refers to mutual accessibility. In this definition, everyone on the team feels that all team members are accessible for answers and support. Rice notes,

> Sometimes there are project teams where the project team leader in the job site trailer has their office door closed all the time and is not available, but the second they need something, they don't mind going out to their project engineer's desk and expecting them to drop everything to give them any answers they need. However, when the engineer has a question or needs answers, they feel like they can't disturb the project team leader. That is an example where mutual accessibility does not exist. Therefore, Collaborator Safety does not exist.

Mutual accessibility is also evident in the data indicating response time to emails and text messages.

As the task force reviewed their data, breaking down the information from all of their interviews of a couple of hundred team members, they realized that it was difficult to find teams where all four safety factors existed. When they did, the team member experience metrics were higher, and the teams were performing well. This was useful and actionable data.

Recommendations

Ultimately, the task force put together several recommendations and developed some new tools for team development. They launched pilot "Norming Sessions" for new project teams to see if these tools might be useful for building teams or predicting team performance. Each task force team member led at least three of these pilot programs.

Nancy Rice led three pilot Norming Sessions for projects from the Houston office, including projects for a university indoor sports center, a baseball training facility, and a tourist experience hotel. The program opened with activities shaped to better understand each project team member's personality profile and group conversations about the differences, perceptions, and preferences of various team members.

The event then progressed to "Norming" sessions so team members could learn how best to work together. Participants answered twenty-six standard questions related to typical daily job site scenarios. Each attendee reflected on personal working style preferences and their expectations of others. Rice remembered, "Posing those 26 questions to all team members—letting them all cast their votes and then identifying where there were discrepancies allowed a facilitated a dialogue around what individuals' expectations were [related to performance." Once discussed and negotiated among team members, shared expectations of the team became a charter or a contract for all team members for typical site behaviors and processes.

Change the People or Change the People

The task force members and the MSVN leaders attending the sessions were generally satisfied with the results. The teams moved rapidly through tougher stages of team building within the workshop environment and immediately after than teams typically did. It was the "Norming" sessions during the program that were the most revealing. These sessions often uncovered a lack of alignment of one or more persons on the team, mostly related to working preferences. The task force was careful to support intellectual diversity within teams, but there is a difference between intellectual diversity and working alignment. Rice reflected,

> We absolutely encourage intellectual diversity to avoid group think, which is why we include, Challenger safety, and Collaborator safety in the four buckets. That's not what we're talking about [with working alignment], on these day-to-day behaviors – which can be very specific and tactical because they're on the job site. Something as simple as whether agendas and meeting minutes are needed for every meeting could create conflict. One of the items uncovering critical differences in alignment was work hours. For example, the job site typically opens at 5 am every day. One team member might demand every team member arrive at 5 am for bend and stretch without exceptions and regardless of the team member's role. Other team members might react to that demand saying, wait a minute, you know I can't be here by 5 am because I do the night closeout, so I'm here at 7 pm. Why do I have to be here at 5 am, if I'm here till 7 pm and you guys get to leave at 3 pm?

Rice further observed,

> Sometimes these behavioral or procedural issues resulted in an impasse not resolved during the workshop. There were sessions where leadership was able to immediately identify there were going to be issues with the team. Based on the Norming session, it became very clear [on a couple teams there were] one or two team members whose expectations were always completely different from the rest of the team. And even though they ultimately got agreement on the rules for the project team through the facilitation process [during the session], the task force and leadership walked away knowing that one person's expectations are innately different than the rest of the team.

The leadership had not felt comfortable making team composition decisions based on the pilot program observations. However, they did feel that creating contingency plans for some teams was a good idea. When the anticipated problems from the "Norming" exercise emerged, the leadership team was prepared to move quickly, changing out isolated team members for members that fit better. The task force tracked the data from the "Norming" exercises. It determined that 80 percent of the time when the "Norming" exercise revealed isolated team members, it meant that the team member must be coached proactively to engage positively with the rest of the team. Without this coaching, the team member would continue to have behavioral and procedural differences, leading to team dysfunction. The task force was not yet ready to make a firm recommendation, but they kept an eye on this dynamic and encouraged team leaders to coach their isolated members.

Outcomes

The pilot programs were deemed a success at MSVN. The project teams going through the pilot programs ramped up more quickly into the performing phases of teaming. The "Norming" exercises resulted in team members feeling safer, bringing up topics that they typically wouldn't have talked about as early in the process. One particular team, however, never got to "Norming." Rice reported,

> Part of the team was behaving one way and the outlier was behaving another way. There was frustration around not being able to come together which grew quickly, came to a head, and then was able to be resolved. Instead of stringing along for the duration of the project, a year or more, it came to a head in only a few months.

The task force credited the "Norming" exercise with accelerating this process and with faster resolution.

The task force also created a new tool, the Discovery and Safe Tool, now used by all project teams at MSVN for team building. The tool was also rolled out to all MSVN's business groups in 2019. The Team Norming sessions are now a standard program offering used at the discretion of each project team, depending on the size of the team and the duration of the team engagement. The project teams work with the office HR and group leadership to put the programs together as needed. Nancy Rice advocates for both the tool and the Norming session to be used in concert for the best results.

These tools and sessions have been used on teams that have included architects and clients with similar success, including sports projects and projects for technology companies. Rice remembered one sports project where the design team didn't want to participate in the process. It was fairly clear that they didn't want a collaborative process, an important red flag. The tool and program provide essential insights to potential problem areas in creating psychological safety and identifying team members that may not be great collaborators.

Finally, the MSVN leadership had the answers they sought regarding high-performing teams. The research from the task force had given them critical insights regarding the elements that make teams successful. MSVN leaders' traditional claim "We Build Anything" is more true than ever. Now they can build great teams as well.

High-Performing Collaborative Teams

3

You Can't Fake it

When the USA Women's team won the 1999 World Cup Final against China in the Pasadena Rose Bowl, and Brandi Chastain famously ripped off her shirt in celebration, women's sports were changed forever. According to Jere Longman in *The Girls of Summer*, few media outlets were interested in the event, and almost no major newspapers sent reporters. Yet, the Rose Bowl was packed with 90,000 spectators that day just as all the venues across the United States had been during all the events leading up to the final. An additional 40 million people watched the event on American television. At one time, those in sports analysis (considered an oxymoron by some) felt that women's sports, and specifically women's soccer, would never draw a paid following. This team, this American women's soccer team, changed all that, a team that, upon reflection, set new standards of collaboration.[1]

According to Longman, "the United States reached the final [in 1999] by playing an attractive, selfless, joyous, attacking style that, at a slower speed, brought a certain clarity to its game, which celebrated rudimentary brilliance and a reliance on the group rather than the individual."[2]

This team was well tested. They had already won the 1991 Women's World Cup, a bronze medal in the 1995 World Cup, and a gold medal for the 1996 Summer Olympics. "They played fiercely and powerfully, and they were so well-conditioned that they would not surrender a goal in the second half of any of their six games in the 1999 World Cup."[3] Seven of the twenty players on the 1999 team had been playing together for eleven years. Many of the players were close friends, sharing vacations and family events. They shared heartbreak and success. They supported each other during health and personal challenges and, according to Longman, even babysat for each other's children. Again, according to Longman, "When one player went on a date, they all went on a date. When one had dental work, they all had dental work. 'I Will Have Two Fillings' became the password for cooperation, dependability, solidarity, consensus. Given a free night several days before the final, the players chose to spend it, not with their families, but with each other in a Mexican restaurant."[4]

The USA Women's team was the ultimate high-performing collaborative team. When they faced the Chinese team in the Rose Bowl in 1999, they knew each other intimately. They trusted each other. They understood each other's strengths and weaknesses. They communicated elegantly and efficiently. They held each other accountable. The players were amazingly competitive but mutually reliant, and they knew how to bravely readjust when things did not go according to plan. You cannot fake true collaboration, and true high-performing collaboration is critical to creating a culture of predictable outcomes.

The Need for Teams

In July of 2019, Sid Sanders sat down with Hannah Chenoweth to talk about the book he wrote with Chuck Thomsen, *Program Management 2.0*. Sid and Chuck have many, many years of experience between them and the book is an enormous tome of hundreds of pages of valuable insights. But, when Sid was asked by Ms. Chenoweth, "What is the biggest takeaway you'd like readers to have from your book," Sid responded,

> The biggest overarching message is that the industry has and will continue to become more integrated. The line between designers and builders will become more and more blurred, which is a great thing. I came of age as a young architect in the early 80s and the world was starkly divided: there were those who drew designs and had the final authority and those who executed those instructions. That's not an effective way to build today. It needs to be a more collaborative and innovative world.[5]

Yes, we need teams. The days of the single practitioner have passed. The challenges of the built environment are too complex and there is far too much at stake. Even the one-person architectural firm has to know how to work as part of a team to get even the smallest projects completed in the twenty-first century. More commonly, we are faced with projects that rely upon dozens of team members during design and hundreds during construction. Working on teams is no longer a choice and working on teams collaboratively is a required skill set of the twenty-first-century design and construction industries.

Randy Deutsch expressed it in even more compelling fashion in his book *Convergence*,

> Architects and other design professionals today are expected to design and construct in a manner that uses fewer resources, while still innovating, adding value, and reducing waste. Deliverables have to take less time and cost less money to produce, while not compromising on quality—expectations that many feel are unrealistic at best, and often result in a negative impact on outcomes, working relationships, and experiences. Old paradigms such as 'Quality, speed, and price: pick any two' no longer apply. Owners expect all three—perfect, now, and free—on almost every project.[6]

Design professionals that attempt to deliver all this on their own will fail.

To deliver better and faster with less, we in the highly complex design and construction industry will have to be more nimble and more resilient. In addition, we must create a culture of innovation, similar to challenges documented by General Stanley McChrystal in his book *Team of Teams,* as he faced terrorist groups fighting a new kind of war.

> In the course of this fight, we had to unlearn a great deal of what we thought we knew about how war—and the world—worked. We had to tear down familiar organizational structures and rebuild them along completely different lines, swapping our sturdy architecture for organic fluidity, because it was the only way to confront a rising tide of complex threats [...] We looked at the behaviors of our smallest units and found ways to extend them to an organization of thousands, spread across three continents. We became what we called "a team of teams": a large command that captured at scale the traits of agility normally limited to small teams.[7]

The general understood that high-performing teams could address complex problems within complex organizations, and could be resilient and adaptable. His high-performing collaborative teams had to innovate to be ready for disruption. The design and construction industries need these kinds of teams, teams that can adapt, be innovative, and bravely readjust.

It Ain't Easy

Many books have been written about the subject of collaboration and teaming. One of the most interesting examples is Amy Edmondson and Susan Reynolds' *Building the Future*, a book that argues the need for super teams to support megaprojects, where the complexities of future challenges will demand extraordinary teaming.

> This is because bringing together diverse elements (technologies, plans, people, or organizations) to create a functioning whole presents countless ways for integration to break down. Teaming across disciplinary and industry boundaries is needed to respond to the spectacular challenges the world faces today, but it requires a new way of working, a new way of thinking, and a new way of being.[8]

Our industry track record is uneven, as Edmondson and Reynolds note, "The planners, architects, and builders who collectively translate ideas into concrete reality have long played crucial roles in how any large urban project turns out. Sometimes their interdependence plays out collaboratively, other times contentiously."[9] Sanders and Thomsen reiterated this very simply in *Program Management 2.0*, "Collaboration is something that we need to understand, manage and improve."[10]

Our ability to collaborate is tangibly and visibly inconsistent, not only to other team members but to those outside our industries. The good news is this is a challenge that can be addressed. To quote Sanders and Thomsen again,

> Yes, some people are naturally endowed with these interpersonal skills, and some people are just plain difficult. But teams can improve. We need to understand the basic mechanics of interpersonal relationships, the predictable behavior of groups on project teams and the forces that influence the actions of men and women who work together during periods of stress. We need to learn how to create and manage collaboration—to understand the architecture of teamwork. And these things we can study and learn about.[11]

Expressing the need to understand team behavior and dynamics is the first step to better collaboration.

Now What?

Team Meeting #1

The Real Deal about Teams

A few months ago, I laid aside six proposals from different construction companies, shaking my head after a careful review. I had been assisting an owner team with architect and contractor selection to determine which teams would help them be most successful within their strict state policy standards. Proposals, no matter how good the RFP, are not particularly helpful when selecting a team member because proposals are often a compilation of prewritten materials and are usually not written by the key project team members. Proposals are most often written by marketing departments.

I wondered as I read the proposals submitted for a medium-sized ($20 million) renovation project, exactly who had written these particular proposals and, in some cases, what the authors were thinking. Only one submission revealed the smallest understanding of the real qualities of collaboration even though the RFP had placed great emphasis on the subject. In response to a request to provide examples of teaming expertise, one contractor wrote 1) they often provide space in their trailer for the design team, and 2) they bring subcontractors in for a precontract meeting. "Wow," I thought, "that was underwhelming!"

In another competing proposal, from a national-level contractor, boiler-plate inserts implied the company understood the tools and elements of teaming and collaboration. However, short paragraphs, inserted here and there, left the reader wondering if the contractor's heart was genuinely dedicated to collaboration. In the proposal, the contractor offered to manage the design process and the architect's deliverables on behalf of the owner. The offer was veiled, but very clear. The proposal also stated that the selected delivery process, a collaborative Guaranteed Maximum Price (GMP) delivery approach, would be VERY advantageous for the owner. The proposal never mentioned benefits or opportunities of GMP for other team members.

Interestingly, it was the proposal from the youngest team and firm that seemed to get it right. Their response demonstrated they had actively practiced collaboration. They not only used vivid adjectives to describe the teaming process but the proposal also detailed skills that had to be practiced to build relationships and trust. This document described listening skills, problem-solving behaviors, and transparent knowledge sharing as practices that define the teaming relationship. The proposal also suggested the owner and designer participate in setting a collective vision. This proposal provided case studies illustrating how their teaming practices worked, and the case studies that were cited included members of the proposed project team. Best of all, the proposal even spoke to team values.

Where high-performing collaboration is concerned, you cannot fake it, even in a written proposal. To experienced professionals that know what effective collaboration is, and understand how successful it can be, the evidence of practicing the skills of collaboration is obvious. No one can pretend to understand a process that they have never practiced. Real collaboration, authentic high-performing team experience, is not just having participated on a team. In school and at work, we are often thrown on teams and told to get on with it. Many of us grow up thinking we have, therefore, collaborated. Some, as a result, have a bad attitude about teams. They have had unfortunate and disappointing team experiences because they had no idea how to shape a great team experience.

Without training, without guidance, teams will often fail, and those on the team will have a bad experience. No wonder there are so many people in the world that would rather work alone than have to suffer another team experience. Yet those who have the experience of working on great teams and, more importantly, understand why those teams worked well together, prefer teams, and seek teams to address difficult challenges. True high-performing collaboration is created by practicing a set of skills every single day to make your team successful.

Years ago, I was inspired by a book by Leonard Berry and Kent Seltman titled *Management Lessons from the Mayo Clinic*. At the time, I was studying higher education management, but these particular lessons apply beautifully to the complex world of the design and construction industries:

> Collaboration, cooperation, and coordination are the three dynamics supporting the practice of team medicine at Mayo Clinic. These dynamics drive the delivery of personalized care for patients, although staff members care for thousands of patients each day. Individual staff members—from physician to custodian—become active team players to serve patients' needs because treating complex illnesses requires the diverse expertise available from all personnel and the supporting infrastructure. To work at Mayo is to be on the team.[12]

The starting assumption at Mayo is that you are part of the team, and when you are part of their team, no matter what role you play, you adopt the guiding principles, in their case, collaboration, cooperation, and coordination. These would be great starting principles for any team. Just don't fake it.

Stop Playing Nice and Collaborate—The Skillset

April Heinrichs "intimidated everyone," according to the first coach of the USA Women's soccer team, Anson Dorrance. He made Heinrichs the team's first captain in 1986 because she pushed hard and competed hard every day. When Brandi Chastain joined the team in 1988, she confided Heinrich's behavior "scared the hell out of me."[13] Traditionally, in women's soccer, that level of aggressiveness was not well received and came with a personal cost.

"I've spent my life trying to develop this kind of player," Dorrance said. "And to protect them from the sociology of their culture, protect them from their teammates who want them to pass the ball, from the parents of their teammates who are screaming that they're selfish, from everything in the world that tells them a pass is better than a shot, which is a crock. A shot is better than a pass. And it is only the really powerful personalities who are capable of taking the shot all the time, despite the criticism and some watered-down media image of what team success is all about.[14] The thing I admired about April is, she wanted to be liked, wanted to be on a team that got along, but she wouldn't sacrifice her level of excellence to be like everyone else, wonderfully mediocre. We took her mentality and framed the culture of the national team around her."[15]

Making Teams Work

Collaboration is not easy. High-performing collaboration is an everyday job where active collaborators practice a very specific set of skills. It takes discipline. It takes care. It takes patience, and it takes buy-in from all team members.

So, what makes a team work? Many people believe that when you are part of a team, the most important thing to do is to be nice and to just get along. Nothing could be more wrong or dysfunctional for a team. As we reflected in the previous chapter on leadership, a sophisticated-caring leader must create an environment where team members feel safe to speak out and to tell the truth. Truth is critical for excellence and for innovation. Creating this environment takes hard work from team members as well as leaders. When team members choose to be nice rather than collaboratively candid, they are probably not speaking about uncomfortable truths.

Margaret Heffernan illustrated this dynamic perfectly in a 2015 article, in which she said, "in organizations with high degrees of social capital, disagreement doesn't feel dangerous, it is taken as a sign that you care; the best thinking partners don't confirm your opinions but build on them."[16]

This dynamic doesn't give anyone permission to be mean or destructive, but it does require communication, accountability, and excellence to be a priority. This behavior is easier said than done and only accomplished through hard collaborative team-building work.

This skill is only one of many necessary for effective team building. After forty years of leading teams and team building, I have created my own list of these skills, a combination of learned experience and personal research. This list gets tweaked regularly as I learn a new lesson or as I encounter research that helps illuminate behavior I have witnessed. Most of this list is common sense, and when you stop and think about these skills, you will think, well, I already knew that. So, if you knew that, why aren't you practicing that skill with your team every day?

Growing Pains

Before we get to my list of collaborative skills, you need to understand that teams do not come together fully formed or fully functioning, ever. Teams have unique personalities and cycles. They change over time. Each team, even teams with members possessing excellent team skills, will mature as a team at a unique speed and manner. When I started my MBA at the University of Miami in 1998, the faculty spoke to us on the first day of orientation about building teams. They explained that all teams go through a development process that includes several stages, "Forming, Storming, Norming, and Performing." I later discovered that this was a team development model first introduced in 1965 by Bruce Tuckman to explain the evolution of team behavior.[17] Since this model was first proposed, a number of researchers have argued that these are not all the correct stages or that different teams will develop differently based on composition. Alternatively, some researchers claim that team maturity is unpredictable.[18]

Regardless of the pace or process of team development, what you need to know is that the team you start with is not the team you will end with. Your team will change and mature with thoughtful nurturing and the constant practice of team skills. This maturation is documented in the research of Marks, Mathieu, and Zaccaro, who studied the direct relationship of defined processes and values to team effectiveness.[19]

The research from Renee Cheng and Katy Dale also reinforced the benefits of specific collaborative skills in the development of team performance and beneficial outcomes within the design and construction industries. Cheng and Dale documented within their *Teams Matter* study in 2016 the team skill relationships versus sustainability outcomes on the American Recovery and Reinvestment Act (ARRA)-funded projects overseen by the General Services Administration (GSA).[20] This study demonstrated that outcomes are worth the growing pains of team building.

The List

My list, which I ironically title *Collaboration Made Easy*, is part **road map** and part **advice column**. The road map items are steps to take and boxes to check off. The advice column part aspect shows up as items on the list that often do not come naturally to team members and, in fact, go against our instincts. Remember, much of what we have been taught, and most of our instincts relate to personal survival, individual gratification, and safety. All that instinct has to be tossed out of the window when you want to be an effective member of a high-performing collaborative team. So, here is my list:

Collaboration Made Easy

1 Talk About Team Rules, Roles, and Core Values First
   * Define norms and consequences
   * Determine the rules of engagement and cross-boundary processes
2 Define Goals and PROBLEMS Together as a Team
   * Amp up All communications
3 Communicate Effectively and Transparently
   * Include shared language, budget, and problem-solving
   * Apply this rule to the owner and YOU too!
4 Don't Be a Hero
   * Ask for what you need to be successful
   * Tell the truth as soon as possible
5 Take Responsibility and Apologize for Missteps
   * Forgive others too!
   * Deal with conflict quickly and thoughtfully
6 Be Accountable and Hold Others Accountable
   * Practice accountability
   * Respect commitments, and
   * Measure what's important
7 Stay Curious
   * Ask questions without fear
   * Recognize other ideas and disciplines can inform your ideas and spark new ones,
   * So be curious about how others think
   * Seek technologies and processes to support team needs
8 Best Idea Wins
   * It doesn't matter where an idea comes from!
   * Leverage diversity
9 Bravely Readjust…
   * It's an iterative and imperfect process
   * Be resilient and adaptable
10 Succeed and Fail WITH the Team Whatever Your Role
   * Be a member, a leader, a service provider
   * Collaboration Made Easy

DO THIS EVERY SINGLE DAY

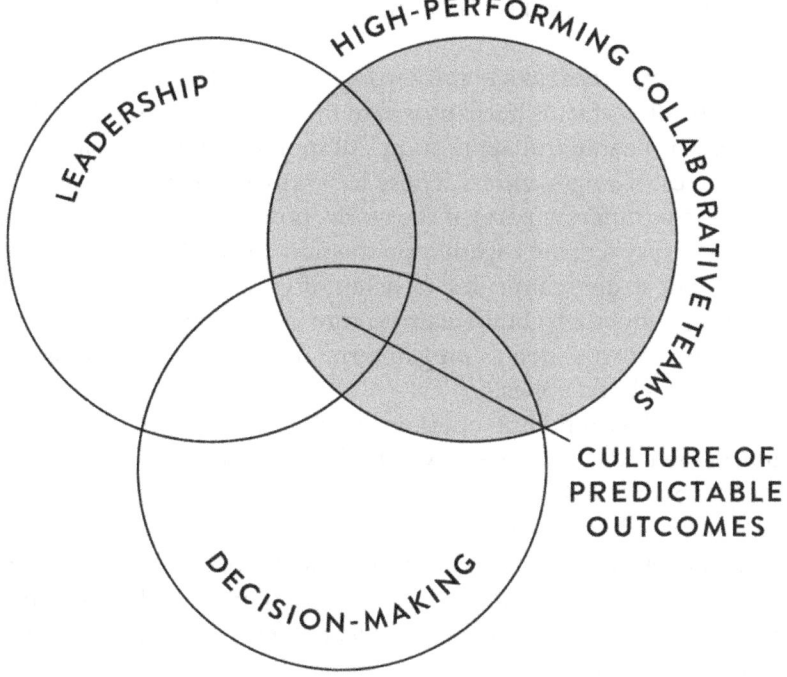

THE SKILLSET FOR TEAMS
INCLUDES:

COLLABORATION MADE EASY

1. Talk About Team Rules, Roles, and Core
   Values First
2. Define Goals and PROBLEMS Together as
   a Team
3. Communicate Effectively and Transparently
4. Don't Be A Hero
5. Take Responsibility and Apologize for
   Missteps
6. Be Accountable and Hold Others
   Accountable
7. Stay Curious
8. Best Idea Wins
9. Bravely Readjust...
10. Succeed and Fail WITH the Team Whatever
    Your Role

DO THIS EVERY SINGLE DAY

Figure 3.1 Collaboration skillset

1 Talk about Team Rules, Roles, and Core Values First

Nothing is as important as a team starting on the right foot, which means in sync. This healthy start is perhaps where the sports metaphors in this book have the most meaningful application. All sports have rule books providing the basic rules of engagement. These basic rules make it possible for two teams from two different parts of the world, possibly even speaking different languages, to play a game together on the same field. Rules of the game are important, but so are team rules. Coaches often set curfews, practice times, and meeting etiquette to build team culture and discipline. Players set their own rules and norms such as pregame rituals, weekly dinners, or strength competitions to build cohesion.

Each of us has different operating styles and different values. These differences are under stress when you build diverse teams. These differences in style and work practices can quickly undermine team building by creating tension and frustration. Amazingly, much stress can be avoided by having a conversation about team norms at the beginning of the engagement. Robert Galford's book *Simple Sabotage* guarantees that committee or teamwork will be sabotaged without clear roles, goals, plans, or processes for discussion and resolution.[21]

Committing to a conversation setting team rules of engagement means teams talk about the easy stuff and the hard stuff. The easy stuff includes performance expectations and consequences. For example, some teams have set consequences for meeting tardiness, such as having to sing a song to the meeting attendees or to buy lunch for the team after two late meeting arrivals. Another team set the expectation that the team would stay positive about goals so they would never say, "I will try…" In this case, every time a team member made this statement instead of "I will…," they were penalized with a beer mug placed next to their name on the meeting notes. Whenever a team member earned six mugs, they had to buy a case of beer for the next team celebration. Your team may wish to discuss expectations and consequences related to making or missing deadlines. Other items in the easy category might include rules about meeting documentation, decision documentation, remote meeting participation, colocation norms, bringing food to meetings, or regular social activities.

The hard stuff includes determining how decisions will be made, how information will be communicated, and how conflicts will be resolved. It can also include rules about meeting etiquette, so that all voices are heard, and the right questions are asked. Some teams have suggested that only one conversation occurs in a meeting at one time. Other teams have put rules in place encouraging all team members' questions and ideas be heard even if the subject is not within an individual's direct area of responsibility.

The beginning of the project is the time to discuss team roles as well. Many misunderstandings emerge related to the definition of role expectations. It may surprise you how many people are willing to be very upfront about their desire to lead or to avoid leadership. It may be that your team would like to rotate roles to share the load better or to support professional growth. Regardless of the outcome, the discussion is important.

The more diverse the team and the more complex the teaming environment, the more complex these conversations about rules of engagement should be. Amy Edmondson addressed this in her book *Teaming to Innovate,* as process guidelines.

> In any complex teaming effort it is important to establish process guidelines that everyone agrees to follow [...] Guidelines are needed for specifying points at which separate teaming activities must come together to coordinate resources and decisions [...] [for an example project this meant an] interface existed when anything touched or crossed a boundary. Regular interface coordination meetings were held to manage physical, functional, contractual, and operational boundaries. Through extensive documentation, the team eliminated mistakes that might otherwise have occurred at these boundaries— saving materials and funds, and sparing headaches.[22]

Diversity of profession makes teaming in the design and construction industries even more challenging. As Edmondson and Harvey observed,

> Most people take the norms and values within their own professions, organizations, or industries for granted, sharing a set of largely unquestioned assumptions [...] Individuals with completely different perspectives face serious communication problems when they interact together. When there is too much distance between team members' knowledge, it is particularly difficult for them to learn from each other because they have trouble recognizing the value of each member's input.[23]

For the design and construction industries this means that we have to be willing to learn and appreciate lexicons and technical processes of our team partners to permit conversations about effective processes.

Equally important is to have a conversation about values at the same time the team is determining roles and rules. This activity is so important,

I have written a chapter dedicated to the subject. See Chapter 5 for a critical discussion on Values Aligned with Goals.

Great teams start collaboration by talking about rules, roles, and core values first and do not forget to

- define norms and consequences, and
- determine the rules of engagement and cross-boundary processes.

2 Define Goals and Problems Together as a Team

As noted in the previous paragraphs, we often come to our teams with different perspectives, expertise, and even language. Yet, even the best, most mature teams have moments where they aren't on the same page. Avoiding misunderstandings and, thereby, rework and frustration is best accomplished by having leveling conversations early and often. It's incredible how often team members do not agree on the definition of the team goals because the team has never actually discussed those goals. This issue applies to the overarching definition of the building project as well as to other smaller goals and problems throughout the project.

In the design and construction industries, our teams can be unimaginably large and complex, as Charles Thomsen noted in *Managing Design*,

> there are so many people. Literally thousands that create little bits of information or service—users, owners, subconsultants. We had 35 subconsultants on one project and the contractor had 75 subcontractors. You've got to get the goals clear, so everybody knows what to do [...] and contribute at the right time.[24]

Remarkably, there is something satisfying and empowering about having the entire team buy into a single goal. Alex Ely commented on this particular dynamic in the new book *Defining Contemporary Professionalism*, "Successful teamwork needs cooperation, loyalty, trust and professionalism: working in unison towards one shared goal. Getting collaboration right is not just about creative problem solving, but it is proven that empowerment improves employee satisfaction."[25]

In the leadership section, I discussed the importance of leadership in providing motivation. Leadership will also be involved in goal setting. However, unless the team carefully discusses the scope, impact, and delivery of goals with leadership, there will be ample opportunity for misalignment and false starts.

A great example comes from a recent movie about a real event in the racing world, *Ford versus Ferrari,* where the lead driver at Le Mans, Ken Miles, was asked to pull back from a multi-lap lead so that three Ford cars could finish first in a dead heat. The assumption by Ford and the racing team was that Miles would still be a winner, but instead, the win went to Bruce McLaren because of a rule technicality. As a result, Ken Miles was also denied being the first and only world Triple Crown Endurance Champion. Ford and the racing team had not clearly defined or agreed upon their shared goals at the start of the race. The consequences were heartbreaking.

Problems facing teams also need definition if teams are to address and resolve these challenges efficiently. Albert Einstein is often quoted as saying, "If I had only one hour to save the world, I would spend fifty-five minutes defining the problem and only five minutes finding the solution." I do not know if he truly said this, but it has always made sense to me because I have seen many a team flounder before carefully retrenching to define a goal or problem.

In *The Owner's Dilemma*, I described a challenge facing the team building the Library Service Center related to some missing underfloor electrical ducts. Weeks were lost before we held a "locked door session" to solve the problem, where we spent the first fifteen minutes of the meeting simply defining the problem. We did not spend time attributing blame. We patiently walked through all the facts. What was there, what was not there. What were the program performance requirements? Only after that process was complete was the team permitted to move on to a discussion of options. The resulting discussion was fast moving and efficient because it was focused on the exact problem to be solved supported by facts.[26]

Strong teams define goals and problems together and

- amp up all communications.

3 Communicate Effectively and Transparently

Would you drive a car wearing a blindfold? Or do you think nothing bad would happen if you drove a car without a working fuel gauge. Perhaps you would recommend that we remove the game clock from football, soccer, and basketball so that the players and coaches find out the game is over only when the whistle blows. Or maybe the umpire in baseball should keep the ball and strike calls a secret until a player is out or walks.

Many industry leaders ask me to advise them about transparency yet reject my recommendation that they create a culture of shared information with their teams. Owners want to hold back information about fees and contingencies. Contractors want to maintain hidden costs. Architects want to control the design story, revealing the complete design only to their best advantage, or until it's too late to change direction. Each of these team members is creating a vehicle destined to crash into an unseen barrier at any moment. This process of hiding knowledge undermines the team's ability to do its job and create predictable outcomes.

Cohen Sharples noted how important transparency is to project success in his essay in *Building (in) the Future*,

> Information flow [...] largely determines the architect's access to the knowledge base of the construction industry. This accessibility is critical throughout the project's development if the design process and the end product are to achieve a high level of efficiency through informed decision making.[27]

Sharples noted in these remarks that knowledge flow is directly related to the delivery process, but what Sharples and many others in our industry often do not understand is that the decision to share information transparently is a decision that can be independent of the delivery process. Although lack of information transparency is most often found, without contractual consequence, in the design, bid, build environment (DBB), a few DBB teams have created processes for generous information flow for problem-solving and decisions. These teams understand something that many teams in our industry do not: sharing information, practicing transparency, no matter what the delivery process, REDUCES RISK. Sharing information reduces the possibility that a team will drive off a cliff, hit a phone pole, or run out of gas. Sharing information reduces the possibility the team will design over budget or count on a secret contingency that isn't there. Sharing information increases the possibility the team will trust the information that is shared, and notify the team when conflicts arise. The more each team member knows, the easier it is to keep other team members from making critical mistakes.

This rule of open team communication applies to the owner, both as contributor and as client. Julia McLoughlin, industry attorney, shared the consequences of not recognizing this principle,

> I am privy to details of many professional standard cases reported to the [RIBA] Institute. It seems there are recurring factors—such as poor communication, issues with budgets or unmet client expectations— that result in an architect being reported for negligence or a breach of professional standards. Many problems stem from a lack of communication between architect and client. Indeed, communication is

listening to a client's needs as well as speaking. Architects need to fully understand a client's requirements. Confirming oral instructions in writing, clearly and concisely, is key.[28]

This comment may seem to be only about documentation issues, but at the core of these issues is a lack of information sharing on both sides. For owners, the entire budget must be shared with the team. There is no such thing as a secret contingency or financial safety net. Everyone knows if there is money to be spent. This is the information age—or, perhaps more accurately stated, the misinformation age. If you are not known as an owner that is being transparent, the team will make up its own story. It is also critical for the owner to make it clear how decisions will be made on a project. Decision-making will be discussed in the next chapter, but a lack of transparency in decision-making is difficult and embarrassing for the project manager who is left without real decision power. Tell the truth so the team can prepare for the real process of decision-making on your project.

Regarding the team members who withhold information because they may desperately want a design element to remain in the project, you are sentencing your project, your owner, and yourself to failure. Participate transparently and build credibility with your owner. If you do so, the owner and the team will support you when it is important to you.

Information is power. Many believe that aphorism means that holding on to information or hoarding information makes individuals personally powerful. In truth, hoarding information on a project weakens the project and increases risk even for the person holding the information. The corollary to "information is power" is that **information unshared is actively destructive**, and information unshared increases risk each day it is hidden. Team members act on what they know to be true. If any team member is hoarding information, another team member is incorrectly acting on misinformation.

Maintaining transparency in sharing project information is a hard decision for team members and leaders, but nothing serves trust-building better than working transparently. Nothing serves problem-solving more than transparency, either. It's crazy to imagine that any team will be effective when operating with only a portion of the information they need. Wouldn't it be amazing to work in an environment when you never again had to have a conversation with anyone that started with the statement "I thought you knew…"

Creating transparency may not be dependent on the delivery process, but that does not mean that you do not need to create processes and structures to support transparency. Planning for transparency is critical to make information sharing easy and part of your daily workflow, something General McChrystal recognized in *Team of Teams*,

> Creating transparency and information sharing at the scale we needed required not only a redesign of our physical plant, but also a rethinking of almost every procedure in our organizational culture. The daily O&I briefing lay at the core of our transformation: this pumped information about the entire scope of our operations out to all members of the Task Force and partner agencies, and also offered everyone the chance to contribute.[29]

Once you commit to transparency, look at your communication processes and meeting processes to reinforce this commitment. Creating transparency does not mean that you simply copy all on every e-mail. Don't bury your team in information. Create "just-in-time information" processes that are systematically transparent. During Renee Cheng's research with the GSA, she found that the thoughtfulness of the communication planning mattered. At State Street South McNamara JCK, the team found, "Effective communication and information sharing is the foundation of successful project delivery. Effective communication is marked by open, straightforward, and direct conversation, imparting reliable information of high quality."[30] In *Teams Matter*, Cheng further observed,

> All GSA projects require a communication plan, but implementation success varied. Whipple [project] leaders prioritized communication of roles and responsibilities to properly direct issues and questions. Furthermore, Whipple leaders proactively scheduled reviews of communication plans at each milestone to accommodate any changes in personnel and/or processes during the various phases of delivery and to re-establish lines of communication.[31]

On these projects, no one had to read minds or guess what someone else needed. Keeping team members informed was part of everyone's responsibility, and it proved to be effective.

Amazing teams communicate effectively and transparently and

- include shared language, budget, and problem solving, and
- apply this rule to the owner and you too!

4 Don't Be a Hero

There is a corollary to collaborative skill number three, communicate effectively and transparently, because of human nature. The corollary is, "Don't Be a Hero." Human nature on teams is an interesting thing. When the going gets tough, even on high-performing teams, individual team members often exhibit behaviors that do not serve the team well even when they mean well. Potentially problematic behaviors include the following:

- I don't want to bother anyone else with my problem.
- I don't want to undermine the team by calling out a team member.
- I made a mistake, but I know I can fix it myself.

I call these behaviors "Playing Hero," and I try to make it clear to my teams that the last thing we need is a hero on any team. We need to share the load, and we need to share all the problems.

Creating a Culture of Predictable Outcomes makes it imperative to make two mantras part of your team culture or rules. The first is, "Tell the Truth as Soon as Possible." It's so important to create an environment where sharing information, even bad news, is valued. Never "kill the messenger." Always be grateful for the information because once you have that information, you can act on it. "Tell the Truth as Soon as Possible" sometimes requires that the team member puts aside their desire to be nice if they are to share information that is key to the success of the team and project. This behavior does not mean the team member is disrespectful. It does mean critical information about progress, quality, cost, or any critical characteristic of the project is shared with the team even if it shines a negative light on another team member. Facts are facts, and the sooner they are shared, the sooner the team can recover.

The second mantra the team should adopt is "Ask for What You Need to Be Successful." So many of us, especially those of us trained as architects in the sometimes masochistic studio environment, hesitate to ask for what we need to get our job done. We do not ask for the resources, the time, or the fee to be successful. Often in the middle of a team process, we are not getting the information or decisions we need, and we do not object because we are afraid of looking bad to the team or to the owner. I have seen so many contractors get themselves in trouble on change orders because they hesitate, often unnecessarily, to present real costs to owners. Some owners make it difficult. But owners are team members, and they need complete change-order information to manage decisions and project costs as early as possible. In this situation, contractors should also consider themselves team members asking for what they need to be successful as soon as possible.

In "The Hard Truth about Innovative Cultures," the authors note collaboration shouldn't be hard. "People who work in a collaborative culture view seeking help from colleagues as natural, regardless of whether providing such help is within their colleagues' formal job descriptions. They have a sense of collective responsibility."[32]

I often tell a story about a $130 million project on the Rice University campus that was delayed because a construction project manager didn't want to complain about the architects. The architects were not delivering the shop drawings on schedule. The project manager was playing the hero, playing the nice guy, while the project was dying. As soon as we uncovered the problem, we were able to solve it, partly a process issue. Once resolved, everyone on the project was ultimately successful.

"Don't Be a Hero" means that every team member must

- ask for what they need to be successful and
- tell the truth as soon as possible.

5 Take Responsibility and Apologize for Missteps

We have been working together across disciplines in the design and construction industries for a long time. It has not always been easy. Edmondson and Reynolds note this in *Building the Future*,

> Teaming across the boundaries between even the interlinked professional domains in the built environment is fraught with conflict. Accounts of the tension between builders and architects abound. Nonetheless these particular professional domains have much in common: shared experiences of undertaking large projects from start to finish, overlapping language, common time frames, and more. They also have a long history of working together across their shared boundary.[33]

One of the hardest collaborative skills to learn and practice every day is to put away the ego and be vulnerable. We all make mistakes, and those mistakes impact our teams. Sometimes those mistakes are related to bad judgment or are the result of a failed innovation. Some mistakes may be related to lack of information, or some may be related to just having a bad day or reacting to the inherent tension related to teaming across discipline boundaries. No matter what the reason for the mistake or misunderstanding, be humble, take responsibility, and apologize, quickly and with sincerity. As a business law professor of mine once said, "Nothing is as meaningful as a heartfelt apology."

As far as team building is concerned, a few heartfelt apologies can be the glue for a high-performing team. "Managing emotions and relationships for collaboration is thus crucial to the success of extreme teaming. Doing so effectively, we argue, sets the stage for overcoming the technical challenges that lie ahead," relates Edmondson in *Extreme Teaming*.[34]

When your team knows anyone can make a mistake, and when the team culture includes the inclination to apologize and forgive mistakes, the team is free to experiment and innovate. Always forgive the mistakes of others quickly so long as that teammate is 1) willing to take responsibility for their mistakes, 2) ready to fix that mistake, and 3) willing to learn from the error. Never, ever blame someone else for your mistake. Robert M. Galford writes that during World War II, the organization that would someday become the CIA created a manual for training members of the European resistance. This manual, called the *Simple Sabotage Field Manual*, guided Allies behind enemy lines to create small but highly effective acts of sabotage that undermined the Nazi organizations by disrupting and demoralizing without being detected.[35] "They were hoping [...] that their enemies would waste plenty of time by deliberately haggling over every word. They knew that even small time-wasters can add up to a significant loss of time and energy."[36] Conflict, no matter how small, can be allowed to undermine teams and waste valuable time.

Value a culture of honest apologies and ready solutions to problems. Help and support anyone willing to admit they screwed up and willing to jump in to correct the problem they created. By the way, a heartfelt apology is not about who is right or who is wrong. It is really about whether you value your team relationships above your ego. Let it go and build your team. Great Teams take responsibility and apologize for missteps and

- forgive others, too!
- deal with conflict quickly and thoughtfully

6 Be Accountable and Hold Others Accountable

I was twelve years old when I first learned profoundly about accountability. Sure, I had learned about chores and homework, but there is one moment, one lesson, that stands out. Mr. Fracht of Great Neck, New York, was my violin teacher. He was a larger-than-life, fifty-something character that would always be lively and engaging when my mother and I would arrive for a lesson. My mother waited upstairs while I took my lesson downstairs with Mr. Fracht at his grand piano. I loved earning Mr. Fracht's approval, and he was generally kind and encouraging. On this particular afternoon, I was concerned. Practicing had taken a back seat to homework and other activities over the last week. I had practiced less than half my usual hour and a half per day, and I knew it would show in my playing. I was nervous as I pulled my violin from its case. I tightened my bow as Mr. Fracht asked me how my practicing went this week. I mumbled a few things about school and tests before I quickly interrupted my answer to tune my strings.

Mr. Fracht sat at the piano and opened the lid with his usual flourish and smile. He always seemed to enjoy the start of every lesson. I groaned inwardly as I warmed up with a couple of scales. "Let's start with the Hungarian Concerto you were to memorize this week," he ordered, and my heart sank. I had not yet mastered the piece. I mumbled another few words about not being comfortable with the concerto yet, and Mr. Fracht smiled again, saying, "But we will give it our best, won't we." He then played the first measures. I pulled the violin to my chin and played only six measures before I missed a note. "Keep going!" Mr. Fracht always said this when I made a mistake, and I desperately tried to remember the next section. For I moment I thought I would and then lost it again. Mr. Fracht stopped and looked at me.

Slowly, I pulled down my violin, looking at Mr. Fracht. I could feel the blood rushing to my face. Then my words poured out, "I had a terrible week at school with two tests and a ton of homework. Mom made me do extra chores this week because I am trying to earn money to go on a school trip, and orchestra practice ran late..."

"Enough!" Mr. Fracht said sharply, although he did not raise his voice. He looked at me a long time with deep disappointment and said quietly, "Excuses do not make the music." He then turned, carefully closing the keyboard lid, rising from the piano, said, "Your lesson is over. Your mother is waiting upstairs." I started to protest but could see in his face that he meant every word. I packed up my violin and left.

I never forgot his words, "Excuses do not make the music." He had held me accountable in the most basic and clear manner imaginable. Mr. Fracht connected commitments and results in a meaningful way, an unforgettable lesson.

Teams thrive on commitments, on accountability, and on achieving results. Teams that do not hold each other accountable cannot effectively move forward to their goals. Lack of accountability exemplifies a lack of discipline and a lack of commitment.

Patrick Lencioni commented on this team-building concern in his book *Five Dysfunctions of a Team,*

> Because of this lack of real commitment and buy-in, team members develop an avoidance of accountability [...] Without committing to a clear plan of action, even the most focused and driven people often hesitate to call their peers on actions and behaviors that seem counterproductive to the good of the team [...] Failure to hold one another accountable creates an environment where [Inattention to Results] can thrive.[37]

In *Simple Sabotage*, Robert Galford holds up committees as an especially bad example of a team functioning without discipline, commitments, or accountability. He discusses how committees go wrong and why committees can be the place where projects go to die,

> For starters, committees represent the possibility for accountability to be lost[38] [...] We can usually sniff out a committee that doesn't work a mile away. Luckily for us all, you can too [...] There's No Deadline— Although certain work has a tight deadline—a product launch, for example—much of the effort of modern organizations seems to drift along at its own pace. Either there's a clear and immediate deadline and everyone knows it—We need to have the business plan in front of the Investment Committee a week from Tuesday—or the deadline is vague—We might have something ready toward the end of the next quarter.[39]

Great teams, high-performing teams are committed to goals and hold each other accountable for their performance. Each member of the team respects the commitments they make and follows through on those commitments. Part of commitments and accountability is carefully defining who is responsible for what. Having the discipline to understand precisely who on the team is responsible for each deliverable empowers those individuals and the team as a whole.

Galford recognized that it isn't always simple to figure out who is responsible and accountable, so he offers the RACI model in *Simple Sabotage*. "Under the RACI model, members of a group can hold one of four roles:

R Responsible
A Accountable
C Consulted
I Informed."[40]

In the RACI model, several people can be responsible for aspects of a task, but only one person is ultimately accountable. Others will be consulted and informed.

It's impossible to hold team members accountable without specific goals and measures directly related to outcomes. I have seen managers in the design and construction industries measure the wrong things for years, often exacerbating complicated team relationships and discouraging communications. Counting change orders and Requests for Information (RFIs) are a perfect example. For great teams, these numbers can be high because the team is working hard to improve the project and is asking great questions proactively. Does a money savings change order count the same as an additive change order? Does an RFI that catches a huge problem count the same as a frivolous RFI? Never discourage communication, curiosity, or team collaboration by measuring items that have nothing to do with your deliverables.

Target Value Design (TVD) is an excellent example of changing a project's dynamic by changing the measurement. Used by major project teams across the country, TVD practices reward creativity when they result in cost savings, but only when savings preserve project goals. In *Maximizing Owner Value through Target Value Design*, Glenn Ballard and Peter Morris note that in TVD, "Allowable Cost ≥ Expected Cost ≥ Target Cost."[41] When breaking down the costs in this manner, it is possible to track savings aimed at the target value goal. The entire team understands the definitions and goals. The team also understands the benefits of meeting that target for the owner and the team. Having these measures in place helps the team members hold each other accountable for specific goals.

Design Pull Schedule Planning, part of Lean planning, is another example of measuring what is important. When the team lays out a specific schedule for decision-making for design at the start of the project, each member of the team becomes responsible for generating timely and accurate decision-ready information for that process. Team members can hold each other accountable for meeting those deadlines.

To some, a culture of accountability, of holding team members accountable for their work, seems at odds in a collaborative culture. Happily, it's a synergistic dynamic as noted by Gary Pisano,

> Accountability and collaboration can be complementary, and accountability can drive collaboration. Consider an organization where you personally will be held accountable for specific decisions. There is no hiding. You own the decisions you make, for better or worse. The last thing you would do is shut yourself off from feedback or from

enlisting the cooperation and collaboration of people inside and outside the organization who can help you.[42]

In my experience, teams should track critical measures each week, and they should apply directly to the project's budget, quality, schedule, or risk. Great teams

- practice accountability,
- respect commitments, and
- measure what's important.

7 Stay Curious

Every team needs curiosity, to be successful. Curiosity is the key to collaborative success as it is to personal success. I have seen teams of brilliant individuals fail because no one wants to ask a question in front of other team members. I have seen teams fail because there is no curiosity about other disciplines on the team. I have seen architects turn off entirely when a discussion turns to information technology. I have seen engineers lose interest when the conversation relates to materials selection, or general contractors disconnect when the presentation relates to interior furniture selection. Sometimes, team members shut down because the conversation is in language that is not understandable to all team members. The downside of this disconnection is that there will be major portions of the project unfamiliar to certain team members and decisions made without all the appropriate or possibly helpful input.

On one of my teams, hundreds of thousands of dollars in plenum-rated cabling was saved by a few critical questions by engaged architects finding alternative pathways. Energy solutions have been dramatically enhanced when engineers committed to developing alternative frameworks to guide materials selection. Contractors have been part of the solution instead of the problem when they thoughtfully evaluate systems furniture solutions reducing subcontracting scope. Staying curious creates opportunities and even innovation.

Ask yourself how many times have you let a word or an acronym go by that you did not understand without asking its definition. Perhaps you did not want to appear uninformed, or you did not want to disrupt the conversation. Whatever your reason, when you do not ask, you leave yourself out of a critical aspect of the discussion and miss an opportunity to understand. You also miss an opportunity to bring others along.

Ninety-nine times out of a hundred, if you don't understand something, at least one other person in the conversation also does not understand. I often hear comments or breaths of relief when I ask clarification questions. The other opportunity you miss when you do not ask a question in a team environment is demonstrating it is okay to ask questions. When I lead teams, I ask lots of questions to model the behavior I want the team to practice. I am always willing to be perceived as the dumbest person in the room on a subject because I am just too darn curious to let my ego keep me quiet. Teams that practice this behavior learn what they need to learn. Remind yourself, you do not need to be an expert on every subject, but you need to understand the language of an issue and the underlying concepts. Be curious. Ask questions and don't judge others willing to ask questions, especially if you are not.

Another way to be curious in a collaborative environment is to be curious about what others bring to the team or about how your ideas might be informed or enhanced by other disciplines. Amy Edmondson calls this Mutual Curiosity in *Teaming to Innovate*,

> Genuine curiosity about what others think, worry about, and aspire to achieve is invaluable for crossing boundaries. By cultivating our own curiosity about what makes others tick, each of us can contribute to creating an environment where it's acceptable to express interest in others' thoughts and feelings.[43]

This kind of curiosity develops respect for the talents, experiences, and training of others.

Randy Deutsch writes about curiosity in his book *Superusers*. Curiosity is one of the X factors Deutsch refers to as a critical talent for the ordinarily narrow-focused industry technology professionals,

> Superusers are naturally curious—they're driven by curiosity—particularly about the world outside of technology. It is not enough for technologists to focus exclusively on mastering technology and tools—for that to be the end of their interests. Some address this by adding "A" for art to STEM subjects, making it STEAM [...] Superusers' interests outside of technology have a positive impact on their work, providing perspective, enhancing their laser focus on a task, making their work more enriching by leveraging alternative reference points.[44]

Finally, do not be satisfied with the status quo. The best teams are curious about new ideas, technologies, and tools to help them do their jobs better.

"What if" is never a frightening question; rather, it is a question that opens opportunities and presents new answers.

For a team to create a culture of curiosity, it must practice four important behaviors:

- ask questions without fear
- recognize that there are other ideas and disciplines that can inform your ideas and spark new ones
- so be curious about how others think
- seek technologies and processes to support team needs

8 Best Idea Wins

When my son was nine years old he had already fallen in love with baseball. He played Little League baseball in a very competitive league in Miami, Florida. Several evenings a week and every Saturday would find us watching practice or cheering at games for the Cubs. The Cubs were very good, with lots of talent, at least for nine-year-old kids. My son, Ian, played shortstop and he pitched for the team. They had a lot of great players, Ryan, John, and Anthony, but it was Marianne that was the most notable talent on the team. She was aggressive, athletic, and knew a great deal about baseball. She also pitched and played second base.

Nine-year-old boys and their fathers aren't always the best at accepting girls as "real players," and, frankly, can be a bit mean. Early in the season, Marianne had a few challenges when it came to being accepted by the team—until they played the Expos early in the season.

Everyone knew the Expos were going to be tough and that they were really good hitters. The Cubs played a good game, however, and they were one run ahead in the late innings. Ian had finished pitching and had returned to shortstop. Marianne was playing second base. The new pitcher, Ryan, was doing fine but a bit wobbly. The Expos had gotten two singles off him and had guys on first and second, although they also had two outs. Marianne was always good at chatting up the runners between plays and she was having a pretty good conversation with Scotty, the lead runner, when Ryan took the mound for the next batter. Marianne unexpectedly called "timeout." She waved all the infielders to the mound and everyone followed.

"Ryan," Marianne whispered once all the boys were surrounding the pitcher, "hand me the ball." Ryan responded with a slightly surprised and disdainful, "You can't pitch, Mary. You're pitching tomorrow. Besides, the coach decides who pitches."

"I don't want to pitch, Ryan!" Marianne hissed. "I want you to sneak the ball into my glove, so nobody sees and then I want you to pretend to get ready to pitch. Can you do that?"

All the boys started objecting at once. "What?!" "That's crazy!" "That's against the rules" "Why do you want to do that?"

"Shhhh!" Marianne hissed again, "We only have a sec because the Blue is coming. I know I can get the runner out. I know the rules. It's called the hidden ball trick. Hurry."

Everyone on the mound was silent for a moment as the umpire came up and tapped Ryan on the shoulder. "Let's go!" he said. Ryan nodded as he slipped the ball into Marianne's glove, then turned to scrub the mound with his foot. Marianne let her glove drop to her side has she jogged back to second base. She made some kind of remark to Scotty, who was standing on the base, which made him laugh. All the players were now back in their infield positions and Scotty, watching Ryan standing to face the pitcher, started to take a lead off the base. Marianne chatted away until she stood between Scotty and the base.

"Scotty," Marianne said brightly as she slapped him on the arm with her glove, "tag! You're out! Blue, I've got the ball!" She held up the ball up, then tossed it to Ryan just as the umpire called, "You're out!" to the poor unsuspecting Scotty left standing between second and third base. Marianne grinned. Ryan whooped. The Cubs bench emptied out, surrounding and embracing their clever teammate. Marianne and her team carried their talents and her baseball knowledge all the way to the championship that year.

One of the characteristics that best distinguishes a great team is when ideas can come from anyone on the team or even outside the team. The team becomes an equal opportunity idea grabber from wherever an idea comes. That means the team members are great listeners and they are open to ideas from any source, leaving prejudice and preconceptions at the door.

The theory of "Best Idea Wins" is especially effective when the team recognizes the benefits and seeks to leverage diversity. "Group Think" is real, and we in the design and construction industries have been suffering from the consequences of this disease for a very long time. Once a team and its leadership have worked hard to create an environment where all members of a team feel they can share ideas candidly, it is time to listen. Professionals in their discipline often feel they are the experts on any problem related to their areas of responsibility and should be consulted first and last. This, of course, discounts the complexity of contemporary construction projects and the fact that so many problems cross disciplinary boundaries.

I have never cared about the source of a good idea. If the plumbing engineer solves a curtain wall design problem, or the architect solves a plenum-rated cable problem, I am thrilled that each was willing to think about challenges outside their normal area of responsibility and interest. When teams are willing to listen to good ideas, even crazy ideas regardless of the source, the project benefits and everyone grows.

Leveraging Diversity

When Canan Yetmen and I wrote *The Owner's Dilemma* in 2010, we carved out an excerpt that *Engineering News-Record* (*ENR*) published soon after the book was published. The excerpt was also published on *ENR's* online site where the comments were published. We were delighted by most of the comments, but one was especially memorable. It went something like this, "When you two little ladies want to put on hard hats and see a real construction site…" I could have written this off as a troll, and, in fact, it never bothered me much. I have gotten quite a few laughs at this guy's expense over the years. However, this writer's attitude is all too evident on many teams. We dismiss people and their opinions because of their lack of experience, their gender, the way they look, their lifestyle choices, or their profession. We stop listening and we are the poorer for it. Part of recognizing there are other ideas and disciplines that can inform your ideas and spark new ones is understanding that team diversity is valuable.

Extraordinary teams embrace cross-disciplinary opportunities and embrace the intellectual diversity that comes with this dynamic. Diversity is healthy for teams, as noted by David Rock and Heidi Grant in *Harvard Business Review*,

> In recent years a body of research has revealed another, more nuanced benefit of workplace diversity: nonhomogenous teams are simply smarter. Working with people who are different from you may challenge your brain to overcome its stale ways of thinking and sharpen its performance.[45]

Rock and Grant go on to inform us that diversity of all kinds makes business sense,

> A 2015 McKinsey report on 366 public companies found that those in the top quartile for ethnic and racial diversity in management were 35% more likely to have financial returns above their industry mean, and those in the top quartile for gender diversity were 15% more likely to have returns above the industry mean.[46]

However, there are those in our industry cautious about diversity, "Well composed teams [...] have a good mix of members, people who are neither so similar to one another that they duplicate one another's resources nor so different that they are unable to communicate or coordinate well."[47] I am, in turn, cautious against these half measures. If you need to have hard conversations about communication and values, do it. Do not be fearful of those different from yourself. As Brené Brown observes,

> People are opting out of vital conversations about diversity and inclusivity because they fear looking wrong, saying something wrong, or being wrong. Choosing our own comfort over hard conversations is the epitome of privilege, and it corrodes trust and moves us away from meaningful and lasting change.[48]

If we genuinely want to seek the best ideas, we need to leverage diversity, and we must be willing to have hard conversations.
Smart project teams know

- it doesn't matter where an idea comes from!
- leverage diversity

9 Bravely Readjust...

Plans and processes aren't everything. Teams that make a plan and then work a plan without realizing complexities and problems may emerge that require adjustments, will fail. The best teams are nimble. They stay focused but also understand they must be ready to readjust when things do not go according to plan because there is much at stake. Chuck Thomsen and Sid Sanders discuss this dynamic in *Program Management 2.0*,

> We have never seen a project that didn't change. Somewhere in the world there must be a project that was executed as planned; we've never seen it. Projects change. It's inevitable. Planning can minimize the effect of a capricious change, but the world's best planner can't eliminate unexpected events.[49]

They go on to assess, "When change happens, timely communication is crucial. Much will be in motion: design teams, manufacturers and construction crews will be working. Products will be flowing through the supply chain to the site."[50]

Bravely readjusting is a mindset for the entire team. It means that everyone understands that things can and will go wrong or may need to change. It means that everyone on the teams accepts that the unexpected can occur and that the team needs to move on from that event as quickly as possible. Assigning blame is not a path forward. Sometimes you must let go of the process. Resiliency is the answer. Assess options and choose a path forward.

General McChrystal observed this in *Team of Teams,*

> Prediction is not the only way to confront threats; developing resilience, learning how to reconfigure to confront the unknown, is a much more effective way to respond to a complex environment. Since the pursuit of efficiency can limit flexibility and resilience, the Task Force would have to pivot away from seeing efficiency as the managerial holy grail. To confront a constantly shifting threat in a complex setting, we would have to pursue adaptability.[51]

Other skills from this list are required to bravely readjust. Maintaining a focus on goals and values is a must. Communication is critical, especially transparent communication. Don't be a hero—tell the truth as soon as possible. Ask for what you need to be successful.
Successful teams remember

- it's an iterative and imperfect process, and
- to be resilient and adaptable.

10 Succeed and Fail WITH the Team Whatever Your Role

In *The Ideal Team Player,* Patrick Lencioni observes,

> Great team players lack excessive ego or concerns about status. They are quick to point out the contributions of others and slow to seek attention for their own. They share credit, emphasize team over self, and define success collectively rather than individually. It is no great surprise, then, that humility is the single greatest and most indispensable attribute of being a team player.[52]

There are many aphorisms relating to individual personalities and teams. One popular one is "There is no 'I' in 'Team.'" I absolutely agree egos are a problem where teams are concerned. Overblown egos and self-involvement result in narrow perspectives, so you only see the impact of decisions on yourself rather than on the broader team or the team goals. When you determine that you are part of a team, you must commit to being part of the team in every respect. That means you succeed, and you fail, with the team. Yes, in the privacy of your team meeting, you can wrestle with issues and concerns, but externally you rise and fall together. This commitment broadens your perspective. You imagine, at every decision point, how that decision impacts the activities of every team member.

In *The Owner's Dilemma*, I wrote about the owner's need to wear multiple hats when they worked with their teams: team member, owner/leader, and service provider.[53] This same frame works for every member of every team. Team members must wear three hats. You are always a team member succeeding and failing with the team, but, at times, you must be a leader of an activity, task, deliverable, or a developing idea. Then, at other times, you will be a service provider, developing content, decision-ready information, or processing invoices for other team members. Each of these roles expands your perspective and deepens your connection to the team.

Members of robust teams know they must

- be a team member, a leader, and a service provider.

Summary

High-performing collaborative teams are only possible if team members are willing to embrace and practice all the skills and lessons of collaboration every single day.

1 Talk about Team Rules, Roles, and Core Values First
2 Define Goals and PROBLEMS Together as a Team
3 Communicate Effectively and Transparently
4 Don't Be a Hero
5 Take Responsibility and Apologize for Missteps
6 Be Accountable and Hold Others Accountable
7 Stay Curious
8 Best Idea Wins
9 Bravely Readjust…
10 Succeed and Fail WITH the Team Whatever Your Role

Notes

1 Jere Longman, *The Girls of Summer: The U.S. Women's Soccer Team and How It Changed the World* (Australia: HarperCollins e-books, 2000).

2 Ibid., 373–375.

3 Ibid., 380–384.

4 Ibid., 386–389.

5 Hannah Chenoweth, "Leadership Series: Sid Sanders, Houston Methodist. We spoke to Sanders about building strong teams and got takeaways from his book, 'Program Management 2.0,'" *Health Spaces*, July 25, 2019, https://info.healthspacesevent.com/blog/leadership-series-sid-sanders-houston-methodist, 9.

6 Randy Deutsch, *Convergence: The Redesign of Design* (West Sussex, UK: Wiley, 2017, Kindle ed.), 469.

7 Stanley McChrystal, Tantum Collins, and David Silverman, *Team of Teams* (New York: Penguin Publishing Group. 2015), 232.

8 Amy C. Edmondson and Susan Salter Reynolds, *Building the Future: Big Teaming for Audacious Innovation* (Oakland, CA: BK/ Berrett-Koehler, 2016), 6.

9 Ibid., 103.

10 Chuck Thomsen and Sid Sanders, *Program Management 2.0: Concepts and Strategies for Managing Building Programs* (rev. ed.) (McLean, VA: CMAA, 2011), 588.

11 Ibid.

12 Leonard L. Berry and Kent D. Seltman, *Management Lessons from the Mayo Clinic: Inside One of the Most Admired Service Organizations* (New York: McGraw Hill, 2008, Kindle ed.), 1353.

13 Jere Longman, *The Girls of Summer: The U.S. Women's Soccer Team and How It Changed the World* (Australia: HarperCollins e-books, 2000), 1172.

14 Ibid., 1197.

15 Ibid., 1185.

16 Margaret Heffernan, "The Secret Ingredient That Makes Some Teams Better Than Others," Ideas.ted.com, 2015, https://ideas.ted.com/the-secret-ingredient-that-makes-some-teams-better-than-others/, 9.

17 Bruce W. Tuckman, "Developmental Sequence in Small Groups," *Psychological Bulletin* 63, no. 6 (1965): 384–399.

18 Amy C. Edmondson and Jean-Francois Harvey, *Extreme Teaming: Lessons in Complex, Cross Sector Leadership* (Binley, UK: Emerald Publishing, 2017, Kindle ed.), 826–852.

19 Michelle A. Marks, John E. Mathieu, and Stephen Zaccaro, "A Temporally Based Framework and Taxonomy of Team Processes," *Academy of Management Review* 26, no. 3 (July 2001), 356.

20 Renée Cheng, "Teams Matter: Lessons from ARRA, GSA Region 5 and the American Recovery and Reinvestment Act," School of Architecture, University of Minnesota, GSA Region 5, 4240 Architecture, May 2015.

21 Robert M. Galford, Bob Frisch, and Cary Grenne, *Simple Sabotage: A Modern Field Manual for Detecting & Rooting Out Everyday Behaviors That Undermine Your Workplace* (New York: HarperCollins, 2015, ebook), 848–882.

22 Amy C. Edmonson, *Teaming to Innovate* (San Francisco: Jossey-Bass, 2013, ebook), 66–67.

23 Amy C. Edmondson and Jean-Francois Harvey, *Extreme Teaming: Lessons in Complex, Cross Sector Leadership* (Binley, UK: Emerald Publishing, 2017, Kindle ed.), 49–50.

24 Michael Alan LeFevre, *Managing Design: Conversations, Project Controls, and Best Practices for Design and Construction Projects* (Hoboken, NJ: Wiley, 2019),13.

25 Alan Jones and Rob Hyde, eds., *Defining Contemporary Professionalism: For Architects in Practice and Education* (London: RIBA, 2019), 90.

26 Barbara White Bryson and Canan Yetmen, *The Owner's Dilemma: Driving Success and Innovation in the Design and Construction Industry* (Atlanta: Ostberg Library of Design Management, Greenway Communications, 2010), 157.

27 Peggy Deamer and Phillip G. Bernstein, eds., *Building (in) the Future: Recasting Labor in Architecture* (New York: Princeton Architectural Press, 2010), 91.

28 Alan Jones and Rob Hyde, eds., *Defining Contemporary Professionalism: For Architects in Practice and Education* (London: RIBA, 2019), 172.

29 Stanley McChrystal, Tantum Collins, and David Silverman, *Team of Teams* (New York: Penguin Publishing Group, 2015), 171.

30 Renée Cheng, "Teams Matter: Lessons from ARRA, GSA Region 5 and the American Recovery and Reinvestment Act," School of Architecture, University of Minnesota, GSA Region 5, 4240 Architecture, May 2015, 23.

31 Ibid., 29.

32 Gary P. Pisano, "The Hard Truth about Innovative Cultures," Harvard Business Review, January – February 2019, https://hbr.org/2019/01/the-hardtruth-about-innovative-cultures?...4a50d3&hctky=11458115&hdpid=c8634bfb-8e7d-4cb7-8c3a-2337818730b9, 11.

33 Amy C. Edmondson and Susan Salter Reynolds, *Building the Future: Big Teaming for Audacious Innovation* (Oakland, CA: BK/Berrett-Koehler, 2016), 103.

34 Amy C. Edmondson and Jean-Francois Harvey, *Extreme Teaming: Lessons in Complex, Cross Sector Leadership* (Binley, UK: Emerald Publishing, 2017, Kindle ed.), 48.

35 Robert M. Galford, Bob Frisch, and Cary Grenne, *Simple Sabotage: A Modern Field Manual for Detecting & Rooting Out Everyday Behaviors That Undermine Your Workplace* (New York: HarperCollins, 2015, ebook), 141.

36 Ibid., 1244.

37 Patrick Lencioni, *The Five Dysfunctions of a Team: A Leadership Fable* (San Francisco: Jossey-Bass, 2002, ebook), 188–190.

38 Robert M. Galford, Bob Frisch, and Cary Grenne, *Simple Sabotage: A Modern Field Manual for Detecting & Rooting Out Everyday Behaviors That Undermine Your Workplace* (New York: HarperCollins, 2015, ebook), 794.

39 Ibid., 882.

40 Ibid., 898.

41 Glenn Ballard and Peter Morris, "Maximizing Owner Value through Target Value Design," *AACE International Transactions*, January 1, 2010, 351.

42 Gary P. Pisano, "The Hard Truth about Innovative Cultures," *Harvard Business Review*, January–February 2019, https://hbr.org/2019/01/the-hard-truth-about-innovative-cultures?...4a50d3&hctky=11458115&hdpid=c8634 bfb-8e7d-4cb7-8c3a-2337818730b9, 12.

43 Amy C. Edmonson, *Teaming to Innovate* (San Francisco: Jossey-Bass, 2013, ebook), 63–64.

44 Randy Deutsch, *Superusers: Design Technology Specialists and the Future of Practice* (Abingdon, Oxon, UK: Routledge, 2019), 15.

45 David Rock and Heidi Grant, "Why Diverse Teams Are Smarter," *Harvard Business Review*, November 4, 2016.

46 Ibid.

47 Erin Carraher, Ryan E. Smith, and Peter Delisle, *Leading Collaborative Architectural Practice* (Hoboken, NJ: Wiley, 2017), 56.

48 Brené Brown, *Dare to Lead: Brave Work. Tough Conversations. Whole Hearts.* (New York: Penguin Random House, 2018), 8.

49 Chuck Thomsen and Sid Sanders, *Program Management 2.0: Concepts and Strategies for Managing Building Programs* (rev. ed.) (McLean, VA: CMAA, 2011), 470.

50 Ibid., 479.

51 Stanley McChrystal, Tantum Collins, and David Silverman, *Team of Teams* (New York: Penguin Publishing Group, 2015), 84.

52 Patrick Lencioni, *The Ideal Team Player: How to Recognize and Cultivate the Three Essential Virtues, a Leadership Fable* (Hoboken, NJ: Jossey-Bass/Wiley, 2016), 2270.

53 Barbara White Bryson and Canan Yetmen, *The Owner's Dilemma: Driving Success and Innovation in the Design and Construction Industry* (Atlanta: Ostberg Library of Design Management, Greenway Communications, 2010), 72–75.

Chapter Supplement—3
Dream Team

Note: Names of participants, firms, and institutions have been changed at the request of the participants.

The past was sitting squarely in the path of the future. The president of the influential yet unassuming southern university, Southern States University (SSU), could see that quite clearly. He knew hard choices would have to be made, but they would also have to be strategic choices.

The university, not the state's flagship but nonetheless essential and growing, represented more than 30,000 of the state's best students and more than $250 million in sponsored research each year. Its Division I athletics had not won many national championships, but students and alumni passionately supported its programs. The institution's medical school was crucial for supporting the state's health care needs and was a key to the university's future. The future of the medical school was the most critical challenge facing the new president, a very public test for his newly assembled administration. To move forward, to allow the medical school to grow, SSU must metaphorically and, possibly, literally demolish the past and find a way for the local community to accept that reality.

The university had long been integrated into the surrounding City of Magnolia, a historical Civil War city with a population of 200,000. The downtown district of the city had seen a few hundred years of change and dozens of economic downturns in its history. Most recently, it was enjoying a bit of resurgence, especially near the university's hospital. The future of the SSU hospital and the medical school were at a crossroads. Both needed new facilities, yet greenfield property was not available in the area. The most promising properties for expansion or renewal contained historic buildings. Relocation to another part of the city was not an option. After a great deal of study, it was clear that two sites were going to be the target of the university for expansion, providing much-needed new facilities. One site contained a much-loved historic high-rise hospital that defined the local skyline. A smaller building occupied the other site but one that was held in equal affection by the community.

The SSU president determined that the best course of action would be to start with the demolition of the smaller building while committing to the community that it would be replaced by a building of equal or greater stature as an iconic structure. This building, which would house the new medical school, would be designed by a great architect, and great design would be a priority. In the president's mind, the new building would be a gift to the community from the university. The City of Magnolia had a reputation for historic architecture dating back to the Revolutionary War. The city did not, however, have a reputation for high-rise buildings designed by world-class architects. The president's vision won the day, and the project would proceed. However, the city residents would be watching critically to ensure that high-quality design was indeed a value that would drive the project.

False Start

The associate vice president of SSU Facilities, Burt Machado, was extremely excited. The prospect of working with a world-class architect was an opportunity he had never had before. In fact, Burt rarely became involved with architect selection processes. He usually left that to the director of planning and design, Darleen McCoy. Darleen was an architect, and it was her job to drive the design portion of every project, just as it was Jason Perez's job as director of construction management to select contractors and to manage the construction process. But this project was too important. This project was special to the president. Burt, an engineer by training, wanted to be involved in every step and would make sure he was directly involved in the selection and the negotiations with these famous architects. He would make absolutely sure nothing would go wrong.

Burt enjoyed the architecture selection process immensely. He made it clear to all the short-listed teams that he was the most influential decision-maker in the process, even though the state's selection process was clearly defined and expertly managed by Darleen McCoy. Burt surreptitiously met one of the firm principals, the internationally renowned George F. Atlas, for a drink when he was in New York. He so enjoyed being seen with this famous architect that he also accepted the renowned architect's dinner invitation. Burt enjoyed the company of the charismatic architect and agreed with George's approach to the project. Burt had made his choice and felt strongly that he would have no problem convincing the rest of the selection committee, most of whom worked for him, to go along with his choice. Burt didn't care how well any of the other two teams did in their interviews.

When the day of the architecture interviews arrived, the members of the selection committee were split on the selection. The two architects on the committee, Darleen and an outside architect from the community, felt the internationally famous architectural firm of PCX had aced their interview and had provided an excellent approach to achieving an affordable yet beautifully designed project. Darleen and the other architect on the committee also liked the architecture firm with which they proposed to joint venture. Dicemore Architects were experienced in medical school design and had worked well with the university in the past.

Burt, of course, felt differently. He did not like the lean, small-framed, soft-spoken lead principal from PCX Architects, Saul Lieberman. He thought of him as weak and not up to the task of delivering an important project like the medical school project. There was just no comparison to the dynamic, savvy George F. Atlas. At the beginning of the selection deliberations, Burt quickly announced that he felt that George F. Atlas Architects was the far superior firm and that he would support no other choice. The other members of the committee supported Burt, and, eventually, the two architects on the committee relented. Darleen was sent off to ask the chosen firm, George F. Atlas, for a fee proposal.

Negotiations with George F. Atlas did not go well. Knowing that Burt had favored them in the selection process, and, given their recent extreme success in the real estate market, the firm demanded a fee totaling 20 percent of the construction cost. This fee significantly exceeded the project budget estimate. Darleen spent a good deal of time and patience working with the firm to revise the proposal, but George was clear with her that the fee was nonnegotiable. Burt was not worried. He believed he had an excellent relationship with the famous architect and took over the negotiations to make sure the university would get the best architect for the project. However, Burt did not fare any better with George. Now that George had the selection in hand, he made it clear to his new friend, Burt, that 20 percent was the fee required to hire his firm. Burt finally admitted defeat and was forced to say goodbye to George F. Atlas.

Burt then turned to Darleen and asked her to negotiate a fee with PCX, the second choice of the committee. Darleen was delighted, and PCX was happy to propose a reasonable fee. Everyone involved was anxious to get down to business in a very collaborative fashion.

Right Team and Right Time

Saul Lieberman quickly realized he wasn't in New York City anymore. The city of Magnolia was historical, in the middle of beautiful countryside, with some nice architecture here and there, but it could certainly get hot and humid—and finding kosher food was pretty much a disaster. But Saul was on a mission. This was a meaningful project. It was a project that was important to SSU, it had a critical purpose, and it was essential to the city to get it right. He had advocated to the firm that they go for this project because Saul felt like it was a "values" fit for the firm. The university had a good reputation of working in a fair manner with both architects and contractors. It wasn't a perfect situation. No state institution was, but Darleen McCoy was known to local architects as a straight talker and she would point out land mines long before anyone would step on them. Saul was also drawn to the president's vision. There was no doubt in anyone's mind what were the principle driving priorities on this project. Excellent design was one of them, and Saul was delighted with that prospect.

Saul knew already, however, that excellent design with the university's budget constraints was going to be a challenge. It would not be possible unless the entire team worked collaboratively. He wanted to be part of a high-performing team. He hoped the rest of the players would be on the same page.

Darleen McCoy was also hoping for a collaborative process. She was relieved that Burt had taken a step back and that PCX Architects had been selected. Darleen felt that Saul was precisely the kind of architect that could work well within the university environment. Darleen also believed her partnership with Jason Perez, the person responsible for the construction process, was going well. The project delivery process was going to be Construction Manager at Risk; therefore, the contractor was selected and brought aboard immediately. This project was also going to be SSU's first experience with design-assist, a process where the subcontractors would be deeply involved in the design. Ballaine Construction, a national firm with an excellent reputation for pre-construction services, was selected and started meeting with the entire team at regular design meetings. Critical subcontractors were also selected. It felt like every individual team member represented the A team from their respective firms, but it also felt as if egos were in check.

Darleen and Jason encouraged some team-building activities early on, including a visit to the architect's office in New York City. In Magnolia and in New York, the team would have dinners together and, sometimes, organize field trips. Once, they visited a historic antebellum plantation and mansion together. At one particularly poignant moment early in the process, the local project manager noticed that Saul was not eating any of the luncheon food offered at the buffet. Without asking, this thoughtful team member recognized that Saul ate a restricted diet of kosher food but had not wanted to make a big deal of his diet needs.

All future lunches included kosher offerings, a consideration that deeply moved Saul, and left an impression on the rest of the team, building trust and loyalty.

Synergy

Soon, the Magnolia team members found they were offering rides to the airport to traveling team members, enjoying the conversations and company so much they were loath to allow them to take Uber or taxis. The team developed its own language for problem identification and troubleshooting. Team members began to trust in the skills and abilities that each team member brought to the project. The architects trusted the university project management staff to negotiate the complicated approval process. Conversely, when the local community members reacted negatively to computer renderings, Saul took the concept drawings home to redraw them by hand for the public presentation. The sketches helped the community understand that the design ideas were still very preliminary and community input would be impactful.

In a milestone moment for the project, finding the project somewhat over budget after design development drawings were complete, the entire team met in the New York offices of PCX to collaborate on project savings opportunities. No one wanted to call the process "value engineering" because the exercise was a creative activity, not a cost-cutting activity. Design quality remained a priority, so the team developed options for reaching the same design goals right there in the PCX offices. The design team developed the ideas, while the construction team estimated the cost. Together the team members discussed the pros and cons and made decisions with everyone in the room. During the exercise, the team found itself further refining some design values, such as sustainability. The state's governor required LEED Silver. However, the team determined that some other resiliency goals could be sacrificed. Durability, however, could not be. The university was not a wealthy institution, so the team held candid conversations within that room about how to invest wisely in this building. That required a trusting environment.

The team met the goals for quality, budget, and schedule. Darleen McCoy would later reflect on the experience as being one of the best of her career. "This project was the 'Dream Team,'" she would say. "We all stepped up when needed. We all respected each other's contributions, and we all enjoyed the process." The project met all the expectations of the president, SSU, and the community, making it one of the university's most successful.

Master-Level Decision-Making

4

The Power of Decisions

It happens nearly every day on a baseball field somewhere in the world, to novice players and, once in a while, to great players. A slow groundball will be hit to the second baseman with a base runner already standing on first. This groundball has the chance of being turned into a double-play ball if the defense successfully throws two base runners out as they approach second and first bases. The problem with a slow ground ball is that it eats up precious seconds. The second baseman has to decide quickly if a double play is possible, and, therefore, throw to second base for the first out, or, he will decide to settle for only one out by throwing to directly first base, abandoning the chance for a double play. All this happens in less than three seconds from the moment the batter hits the ball. Most often, second basemen are known for their decisiveness, making these decisions instantly and effectively. However, every so often a second baseman freezes and cannot decide between second and first; the ball almost gets stuck in the glove. While the player experiences this decision paralysis, both runners reach base safely.

It's About Time

Timing is everything when it comes to decisions in the design and construction industries. Lack of decisions accounts for a large part of the industries' inefficiencies and unpredictability. Amy Edmondson and Susan Reynolds wrote in *Building the Future* that "75% of construction activities add no value,"[1] and it is not hard to imagine that a lack of timely decisions is at the core of the statistic. What owners and many industry stakeholders do not understand is that every decision delayed is a lost opportunity, often at great cost.

In *The Owner's Dilemma* in 2010, I wrote that the "owner's power within a project is absolutely embedded in the delivery of timely decisions."[2] Every stakeholder in the project process understands this to be true. In the fragmented, stressed, and complex environment of the design and construction industries, no teams will be able to move forward effectively without carefully crafting an approach to decision-making within each project. Without a motivated owner focused on making decisions, and without a decision-making process embraced by the project team, decision paralysis is likely to follow.

A culture of disciplined, determined decision-making is also a confidence builder for any team. Thoughtful decision progress feels satisfying and builds team energy. Few characteristics solidify a team as well. We also know that owners known for efficient decision-making may get their choice of desirable team members. Director Arthur Frazier III from Spelman College was quoted in *Managing Design* as saying, "Teams like working with us because we have a 'short decision tree': just our team and the business and finance office."[3]

Yet, decision-making is often the most neglected element of creating a culture of predictable outcomes and, perhaps, the most frustrating missing element for teams that believe, in all other aspects, they are doing a great job at collaboration. Lack of decision discipline and lack of decision rigor can undermine any team, let alone any project.

In 2010, I pointed out that the MacLeamy Curve (Figure 4.1) had demonstrated to the design and construction industries that decisions pushed earlier into design improved effectiveness and cost. I also explained that there is an ideal moment for every decision to be made within a project, as demonstrated in the Decision Power Curve (Figure 4.2) If every decision were made at this perfect moment with perfect knowledge, the value generated for the project would be very high. However, as demonstrated by the Opportunity Cost Curve (Figure 4.3), the opportunity cost of not making decisions at the ideal moment can be much higher, even exponentially higher, to a project and owner. Opportunity Cost is total value lost from NOT making decisions that would have been realized if those decisions been made in a timely fashion. This cost is why making decisions purposefully is so powerful even when some of those decisions turn out to be mistakes. The benefits of making 90 percent of decisions correctly at the

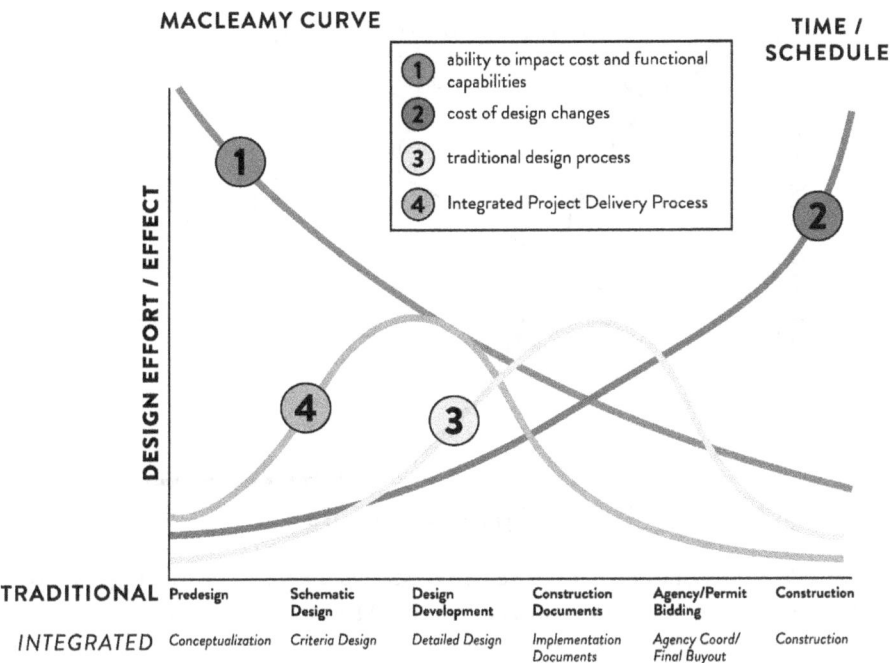

Figure 4.1 The MacLeamy Curve demonstrates the benefits of decisions earlier in the design process through the use of collaborative processes (IPD); as shown in *The Owner's Dilemma*.

right moment outweigh the challenges of correcting mistakes that may (or will) be made.[4]

The barriers to efficient decision-making on design and construction projects are manifest. First, is the overwhelming **desire by some to hold on to decision power** along with the paralyzing **fear of decision risk**. Next, is the **lack of understanding of the consequences of not making timely decisions**, which feeds decision procrastination. A related barrier is the **lack of ability to make decisions stick** once made because stakeholders may not appreciate the cost of changing decisions. **Lack of trust in the decision-making process** creates another barrier, and, finally, the **lack of decision-ready information** available when needed can stop decisions from being made even if the stakeholders are willing.

These barriers must be addressed to achieve successful projects. They can be addressed by creating decision structures that push decision-making to the ideal location within each project and organization, relying on localized experts to inform the decisions and understand the risks of decision delay or changes. In addition, early planning to put decision processes in place on

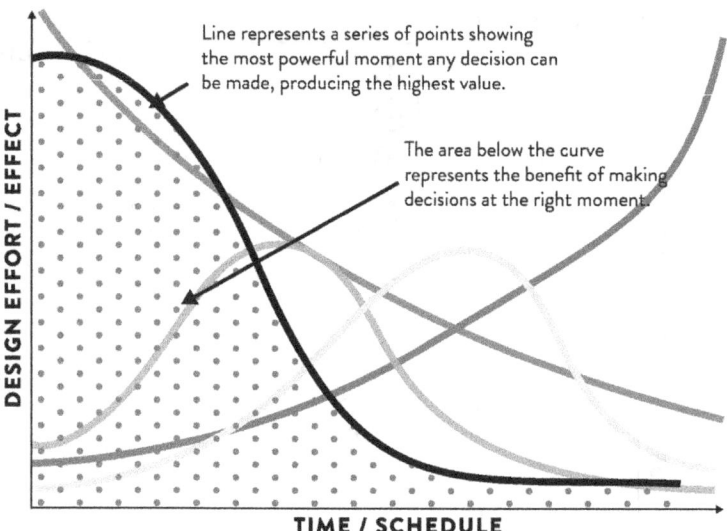

Figure 4.2 Decision Power Curve demonstrates the value of making decisions at the ideal moment; as shown in *The Owner's Dilemma*.

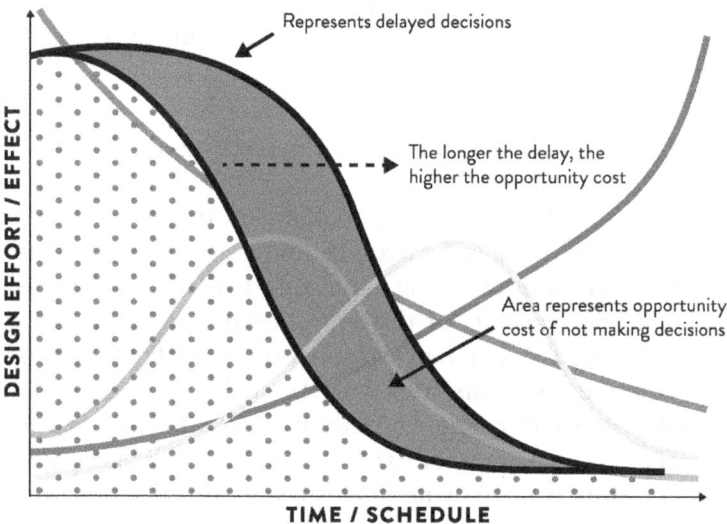

Figure 4.3 Decision "Opportunity Cost" Curve demonstrates the value of making decisions at the ideal moment; as shown in *The Owner's Dilemma*.

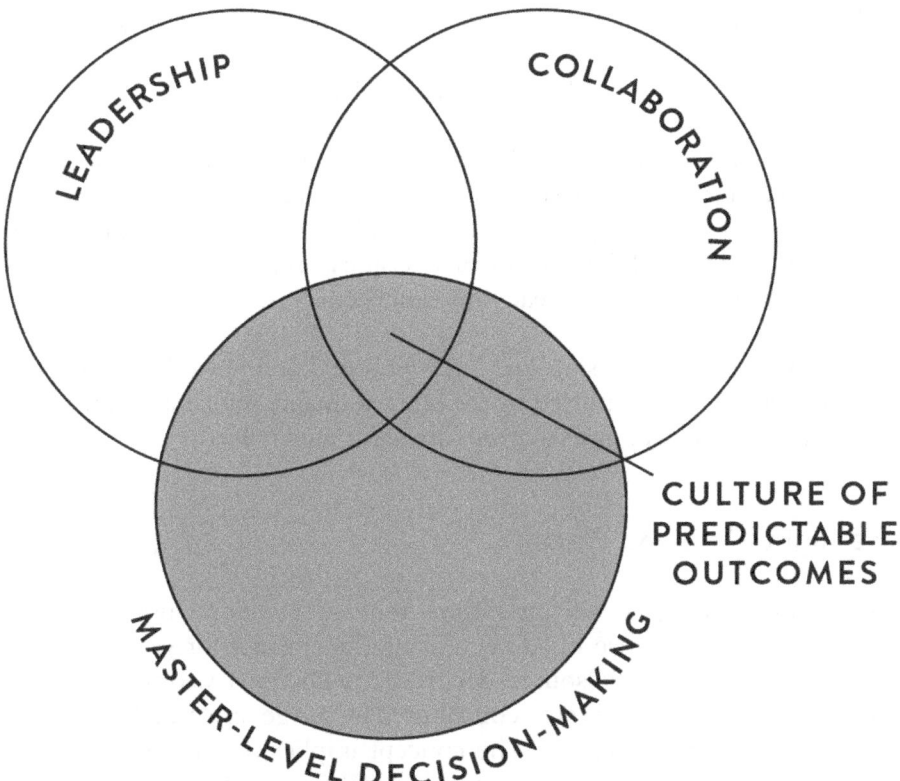

MASTER-LEVEL
DECISION-MAKING REQUIRES:

1. Decision-makers at all levels of projects and organizations to be empowered to make decisions at the right level at the right time;
2. Decision values, principles, and frames to be set early in each project by mapping and scheduling all critical decisions and;
3. The creation of Decision-Ready Information

Figure 4.4 Master-level decision-making requirements

each project that are transparent and anticipate the need for decision-ready information will build trust and accelerate decisions.

Therefore, Master-Level Decision-Making Requires

1 decision-makers at all levels of projects and organizations to be empowered to make decisions at the right level at the right time;
2 decision values, principles, and frames to be set early in each project by mapping and scheduling all critical decisions; and
3 the creation of Decision-Ready Information.

In this environment, decision-making is honored and rewarded, especially when the decision is supported by the highest-quality information, respects stakeholders, and recognizes appropriate associated risk.

The Right Time and Place

I remember I laughed the first time I heard someone use the term "crowdsourcing" because I had been using the concept for years, just not on the Internet. To crowdsource, according to Dictionary.com, means "to utilize (labor, information, etc.) contributed by the general public to (a project), often via the internet."[5] The concept implies that more brains are better when it comes to solving a problem. Well, yes! I think of my teams not as a crowd on the Internet but as a crowd of experts drawn upon to solve problems and make decisions at the right time and place within a project.

The problem with many owners and project team leaders is fourfold. First, they are under the delusion that being in charge means that owners and team leaders make all the decisions. Second, the owner and team leaders want to make decisions because it makes them feel important and powerful. Third, owners and team leaders are anxious about the risk associated with decisions and do not trust their teams to make decisions. Finally, these same leaders believe they are the best ones to decide almost anything.

Unfortunately, every one of these beliefs is misleading and counterproductive. These beliefs lead to decision hoarding and, ultimately, decision paralysis, not to mention decision constipation. Decision-making on design and construction projects is better served through project crowdsourcing. In more specific terms, master-level project decision-making occurs when each decision is made at the perfect level and the perfect moment within a project in partnership with the appropriate stakeholders.

Our discussion of the Decision Power Curve and the "Opportunity Cost" Curve demonstrated why making timely decision-making is compelling and adds value. Therefore, any behavior that inhibits or delays timely decision-making should be avoided. Contemporary design and construction projects of any scale require thousands of decisions to be made every month, sometimes every week. There are too many decisions for any single individual to understand or make in a reasonable or timely manner. In addition, on almost any project, experts on most decisions reside in various places among various stakeholders within the project. These experts understand specific local issues much better than any senior member of the team ever will. Unless the decision impacts a broader set of issues or stakeholders, that decision ideally remains at the expert level. It is there that the best efficiency and value will be generated.

Decision hoarding—when one or more persons on a project claim multiple decisions as their own that would be better made by others—should not be tolerated. Certainly, some decisions must be assigned and owned by specific team members because of the broad or significant impact of those decisions. And, certainly, there should be processes in place to assist when it is not clear who owns a decision. Given these caveats, decision distribution or project crowdsourcing is the most powerful and value-added approach to decision-making on any project.

Your Decisions and My Decisions...An Early Conversation

One of the essential conversations for every team to have before any project starts is how decisions will be made during the project. One of the best ways to get that conversation started is to ask the following questions:

> *What decisions are mine? What decisions are yours? What do we do when there are multiple stakeholders in decisions? What do we do when we aren't sure who owns a decision?*

A hierarchy exists in every organization related to decision-making, just as there is on the baseball field. As noted earlier, the second baseman has only a few moments to decide whether to throw to first or second base. He has no time to ask the coach for guidance. However, a squeeze play, an offensive play where the batter bunts the ball and a player on third starts running during the pitching motion to score at home, requires a decision by the third base coach because it requires coordination between two or more prayers. The manager or head coach prepares the lineup (or batting order) before the game in consultation with assistant coaches. Player recruitment is a set of decisions handled by the head coach or, in professional baseball, the general manager. This hierarchy isn't an arbitrary power grab. Each set of responsibilities for decision-making relates to who best can respond to the timing, expertise, and level of impact or risk of each decision.

Project team members can use a similar filter for assigning decision responsibilities on a design and construction project. Some of the decision responsibilities for significant high-risk decisions may be defined in the contract. Low-risk decisions made quickly and efficiently can be left to project team members without direct leadership involvement. This structure is a significant step forward to empowering project teams.

Defining who owns what decisions helps each team member default to action when they own a decision. Clarifying processes for moments when decisions have multiple stakeholders or when ownership is unclear also provides a path for action. Action is desirable when it comes to decision-making. Remember, no team needs or wants decisions paralysis, the condition where a decision is stalled or when a decision disappears into a black hole, or, worse, decision constipation, a condition where decisions are backing up in the pipeline. Decision constipation is nearly equivalent to a project death sentence. It is difficult to recover from this condition if it goes on too long.

To avoid decision paralysis and decision constipation, each team should have enabling rules around decision-making. In *The Owner's Dilemma*, I shared the simple process I had created for my project managers when I first took over my department at Rice University. This process was simple, clear, and empowering to a group that had been in decision paralysis for some time. The key was to make sure it was safe to make decisions using this process, even when the decision was wrong. These steps were printed on a laminated card and distributed to each team member.

Decision-Making Model

1   *Clearly define the problem*
2   *Identify stakeholders*
3   *Identify and assess risk*
4   *Get the information needed*
5   *Determine if this is your decision…*
6   *Schedule the decision process*
7   *Make the decision… just do it!!!*

If you follow this process and make a decision, even if it is wrong, you will be safe.[6]

Let's dive into these steps for a moment. Defining the problem, step one, is essentially a team-building exercise to get everyone to the same starting position. Often, various stakeholders do not see the problem in the same way or even agree that something is a problem. In addition, problem definition is an opportunity to explain the consequences of not making a decision. The next step, identifying the stakeholders is a quick check to make sure you understand the range of impact of the decision. Step three, identifying and assessing risk is perhaps the most important and most forgotten. Many teams get stuck on decisions that are low risk because they haven't evaluated the risk level of making the decision. Other times, stakeholders have blundered on without recognizing high-risk decisions—with significant consequences.

Step four is all about gathering decision-ready information, a critical process, which will be covered later in this chapter. Step five is a simple concept. Is this your decision to make, or does this decision belong to someone else? Given that, step six allows you to schedule the decision to be made, and step seven represents the discipline to make the decision. As team members, decision-making is part of our job, so we should stay focused on making that happen.

Decision Tools

Pushing decisions to the perfect level and the perfect time in a project requires the right tools, not just discipline. Once your team feels empowered and understands the values, principles, and frames through which decisions can and should be made, it's time to make sure they have the tools to make and push the everyday decisions. Decision flow, the condition of decisions made at the pace the project needs them, is what you want at every level of the project. Decision flow permits RFIs to be answered on time, change orders to be signed before the work is started, and invoices to be paid when contractually required. Decision flow is the ideal state, and the right tools can help your team achieve this design and construction nirvana.

Standard Tools and Processes and Reliable Data

In 2019, Honest Buildings, a company now part of Procore specializing in what is called PropTech, published several blogs targeted at real estate owners. I will discuss PropTech more later in this chapter, but one of these blogs, written by Pauline Nee was titled Cost Tracking Checklist for Real Estate Owners. This blog smartly addressed definitions and risks but also addressed decisions tools as follows:

Get everyone working from the same cost tracking system. Make sure that the entire team is working from one shared document. A reliable construction software platform can keep everybody on the same page, and inform the team of changes in schedules and costs in real time. It can also help avoid delays, quicken invoice approvals, and improve the accuracy of details. Shared software can let every team member track deliverables and materials, and note when payments are due and paid.[7]

A great shared information and tracking system is an important place to start as you build decision-making tools for your team. A shared system increases efficiency, transparency, and shared knowledge. A shared system also builds trust. Almost every decision tool rolled out for a team should be used by every member of the team to be effective.

I spent a significant number of pages in *The Owner's Dilemma* on contingencies, a budget tool that few teams use effectively. Contingencies tend to cause more trouble than they save because most unenlightened team members think of contingency as something to be spent rather than a risk assessment. When appropriately used, contingency calculations at any point in a project will help the team understand how much risk remains in the project and, ultimately, help with project projections. When poorly used, contingencies mask overspending until late in the project when it is too late to respond. Contingency management is one tool that must be understood and used consistently by all team members to be effective. See *The Owner's Dilemma* for more information on project contingencies of all types and how to use these contingencies as risk assessments.[8]

The Owner's Dilemma also addressed other types of decision tools that empower team members to push decisions, helping with decision flow. Two of the most effective are reproduced here:

> **Negative-option e-mails.** An aid to the decision process is the negative-option e-mail, a process by which team members can facilitate decisions they may not have clear authority to make. Project team members can identify sets of decisions that are clearly theirs to make and those that are not. For example, an interior design consultant will likely be charged with approving a substitution for discontinued interior paint color or rubber base molding. To keep the decisions flowing, it is critical that the consultant make these low-risk decisions promptly and efficiently. On the other hand, the consultant cannot decide alone to change all the conference tables to a brand that costs twice as much. That decision would

likely be presented to the steering committee for consideration. Many decisions, however, are a gray area; although not clearly the consultant's purview, neither are they appropriate to be moved up the ladder for approval as it may unnecessarily delay the project. In that case, the consultant or team member can use the negative-option e-mail to make a decision with the short-term option of having that decision vetoed by concerned stakeholders. For example, if the interior design consultant was faced with the need to change the fabric on a significant number of upholstered furnishings but it would not significantly change the design concept or cost, the designer could make the decision, conditioned on the lack of response to a negative-option e-mail. The e-mail may be written as follows:

To: the owner/project manager

Cc: Architect; Contractor

I have been notified by the contractor that the fabric selected and approved by the steering committee to cover all the chair seats in the dining room is no longer available with the protective treatment required by our specification. However, a very similar fabric is available from another manufacturer with the required protective coating. I have attached photographs of the approved fabric and the recommended substitution. I am concerned that the chairs may not be delivered in time for the opening if a substitute is not approved within 48 hours. Renting chairs for the opening will cost approximately $4,000 and an additional $5,000 per week after that. Given the similarity in design and quality between the fabrics and the consequences of moving slowly on this decision, I plan to approve this substitute fabric before the close of business tomorrow afternoon. If you have any concerns with this decision, please notify me no later than early tomorrow afternoon.

In this manner, the designer has broken the decision dam. The decision has been made, but other stakeholders have an opportunity to object if they are uncomfortable with the decision for any reason. Ninety percent of the decisions proposed in a negative-option e-mail either progress without objection or are quickly resolved. Using this tool is far better than a team member sitting on a decision uncertainly or delaying most decisions until the next steering committee meeting.[9]

Locked-door sessions. It is not unusual, even in the most collaborative environments, to have a handful of decisions on a project that just cannot or will not be made. Whether it is the complexity of the decision, the risk associated with it, the unexpected nature of the decision, or simply that competing technologies or goals do not lead to a reasonable solution, these decision dams can pop up like beaver barriers interrupting a vital current. When they arise, it is important that they be quickly identified and resolved. Owners have the ability to hold and facilitate locked-door sessions at which all concerned stakeholders gather in a single room with all the critical information to inform a solution. The gathering's purpose is solely to find an appropriate solution to the problem, and no one leaves until that is accomplished. This meeting should not be used to vent frustrations or to lay blame. It is the owner's job to keep the discussion positive and focused. The facilitation process includes a close examination of all stakeholder goals and foundational information brought to the meeting. Vocabulary and risks are clarified and leveled. Each stakeholder has an opportunity to contribute. With the clear understanding that the meeting will not end until an agreement is reached, the participants are focused and purposeful. This tool has never failed to demolish even the most stubborn decision dam.[10]

Thoughtfully creating tools or even norms for teams that can be routinely used to move decisions through ordinary log jams will be invaluable over time. Other tools and norms might include regular team backlog coffee breaks to determine how to break up backlogs should they exist or be emerging. BIM breakthrough meetings where those knowledgeable in the model share with those less savvy about challenges within the model and/or the design for general brainstorming are also useful. No team member should ever feel like they have no options for raising a concern or leaving a decision stuck in the mud. Figure that out!

Principles/Values and Decision Frames

Once your team has committed to pushing decision-making to the ideal level of your project and organization, you will want to make sure that decisions are made in a manner consistent with the team's values and goals. Consistency in decision-making can be achieved through setting clear principles and values related to decision-making on your project as well as creating decision frames allowing decisions to be made quickly with the right stakeholder participation.

Principles and Values

Every decision flowing through your project should be made reflecting the goals of the project. Most architects will respond that every decision should reflect the contract documents, the plans, and the specifications. However, every person experienced in any design and construction project, whether it be a house or hotel, a strip mall or hospital, understands that there are a million decisions to be made. We cannot rely on the plans and specifications to guide us for them all. Decisions get especially complicated, as David Umstot and Dan Fauchier pointed out in *Lean Project Delivery: Building Championship Project Teams*, because "value is defined by the customer and will be different for each project based on the needs and value definition for those involved in the project."[11]

The authors went on to observe,

> For institutional owners that hold assets for their lifetime, total cost of ownership is important. For commercial owners that have narrow market opportunities to sell their product or beat their competition to the market, project speed will definitely be a value. School projects where students arrive for class in August each year need to meet the schedule date to facilitate this, or miss it by a year.[12]

We will talk more about team values aligning with goals in the next chapter. For now, however, let's think about decision values. At Rice University, I always communicated to my teams that I valued *action over inaction* related to decision-making. That means that I asked team leaders to put in place cultures, processes, and tools encouraging decisions to be made. When a decision could not be made, I wanted leaders to ask why not and when.

Another principle for our teams at Rice was that *we never left a meeting out of budget or out of schedule*. It was the team's responsibility to make a decision how to resolve any emerging budget or schedule problems in the moment, before we left the meeting.

An important project value at Rice University was that *design quality* was very important to each project, so no decision could compromise the design goals of each project. That did not mean we would not substitute materials or save money on some aspects of design. It did mean that design quality, both aesthetically and functionally, was a driving value for every decision. Every decision was tested through that filter.

These examples demonstrate why project teams need decision values and principles to assist. Every project should have its own set of principles and values to be used as decision filters by every team member. These principles and values should be set at the beginning of each project. They include setting the priorities for the building and understanding what can be compromised and what cannot. For example, at a destination facility like Disney, the visitor experience is all important. Therefore, schedule and budget sacrifices are often made to protect and enhance that experience. In a developer-owned speculative office building, the budget and schedule will be sacred. Understanding and accepting those decision principles and values at the start of the project helps the team, drives decisions more efficiently, and reduces frustration.

Decision Frames and Why You Need Them

Decision frames are tools or processes to support decision-making, and they also build trust. Decision frames are organized processes set to make decisions. For example, a space committee might be created at a university to assess and allocate lab and office space, or a firm's compensation committee may review employee salaries and raises.

There are three kinds of trust in team-building relationships, trust in intent, trust in capabilities, and trust in rules and procedures.[13] One of those three, trust in team procedures, can be largely reinforced by building decision frames that are transparent and intuitive to all the project stakeholders. HyperTrack© decision-making and Target Value Design are two examples of action-oriented decision frames. Each process provides a set of rules and processes for decision-making that are consistent and clear. These processes not only drive decisions in a value-added manner but also build team trust.

HyperTrack©

Decision procrastination is the easiest disease to contract in the design and construction industry. We often look for excuses to avoid making decisions. Perhaps we are missing a bit of decision-ready information, so let's push that decision to the next meeting. That order doesn't really have to be completed for another month so, even though we have all the information and buy-in needed, we delay the decision until absolutely needed. A pivotal decision-maker couldn't make it to the meeting, so we push the decision she cared about until she is available.

Each decision to not make a decision can result in cost and capacity challenges later in the project. Decision-making efficiency is a product of a disciplined decision not to permit decision procrastination.

I first learned about the HyperTrack© process from Scott Simpson in 2006 when extreme escalation in construction prices made it imperative that we expedite a large laboratory project at Rice University. HyperTrack©, described by Jim Cramer and Scott Simpson in *The Next Architect: A New Twist on the Future of Design*, became a critical decision strategy for Rice project management for the next $800 million of new construction.[14] Simpson accepts credit for the creation of HyperTrack© in partnership with Pete McCawley, project leader at Amgen, in 1999. The men worked together when the threat of losing one-third in property value in thirty days due to rezoning was an imminent threat. A sum of $30 million was ample incentive to drive the owner, Amgen, and the project team to create a decision frame that allowed a set of permit drawings to be developed in just thirty days (See chapter supplement for the full story).

HyperTrack©, not to be confused with a fast-track management process, provides decision discipline and efficiency to an owner and to a project team. Cramer and Simpson wrote that HyperTrack© "means that all the key decision-makers (owner, architect, contractor, and consulting engineers) are actively and openly engaged in making design decisions as a team from the very first day."[15] As Simpson stated in a recent interview, "Always focus on quality, HyperTrack© is not about speed for the sake of speed."[16] The Rice University team took the Simpson, McCawley HyperTrack© process and customized it for our projects and needs. The key components of Rice's version of HyperTrack© include the following:

- A definition of Decision Values and Principles
- Establishment of clear and consistent decision-making protocols.
- Decision Maps (similar to a design decision pull schedule) created at the start of the project to schedule all major design decisions.
- A schedule all decision meetings developed at the beginning of the project, so all stakeholders know when they need to be available.
- A pre-set agenda for created for every meeting identifying all key decisions to be made.
- Key stakeholders attending all decision meetings or sending representatives prepared to make a decision. Decisions will be made (and not reversed) even if a key decision-maker does not attend.

- Decision meetings to include relevant experts from the project team including owner's stakeholders, consultants, and subcontractors.
- Meetings that always start and end on time. The team will set rules and consequences for meeting behavior (including tardiness) to ensure efficient, inclusive discussion and decision-making.
- All decisions supported by Design-Ready Information carefully defined at the start of the project.

Key advice for HyperTrack©:

- The owner must be deeply committed to this process, must show up, and must follow the rules.
- The decision schedule should be clearly communicated to all team members to build trust and ensure decisions move forward.
- The more inclusive the better in the project meetings. Sometimes the best ideas will come from the most unexpected stakeholders if they feel part of the team and the decisions.
- Don't leave any meeting without being in budget and in schedule.
- All the lessons learned in the Leadership and Collaboration chapters make HyperTrack© more powerful.

Target Value Design

Decision paralysis can sometimes be caused when a team does not know how to make decisions. Overall project budgets and schedules are large targets that are very difficult to aim at without creating a methodology that breaks the targets into manageable understandable pieces. Budget, schedule, and scope are also meaningless without context that can inform decisions. Target Value Design (TVD) is a useful decision frame that has worked for large complex projects from hospitals to airports.

According to an article published by DPR in 2012, "TVD evolved from target costing, a practice made popular in Japan during the economically turbulent 1990s when companies focused on creating high-quality, competitively priced products."[17] However, for the design and construction industries, TVD provides a more specific context.

> Glenn Ballard, research director for the Project Production Systems Laboratory at the University of California (UC), Berkeley, and co-founder of the Lean Construction Institute […] describes [TVD] as a

management practice that 'seeks to make customer constraints drivers of design for the sake of value delivery.'"[18]

In *Integrating Project Delivery*, Digby Christian of Sutter Healthcare was referenced as saying that in TVD "they treated cost as a design constraint, just like any other constraint that had to be considered as the design developed."[19]

Sometimes called Target Value Delivery to communicate that this decision frame extends through the entire project, TVD, according to Umstot and Fauchier,

> is a disciplined management practice that is used throughout project definition, design, detailing, construction, and commissioning to assure that the facility meets the operational needs and values of the users, is delivered within the allowable budget, and promotes innovation throughout the process to increase value and eliminate waste.[20]

Once the owner's value is defined, this decision frame requires a shift in the traditional project team thinking from anticipating expected costs to setting target costs. The fully integrated team works together to develop project solutions because "no single person or discipline has all the knowledge."[21] In TVD, the entire team breaks down the project components and examines those components for value, constructability, and cost. TVD is also focused on eliminating waste in the design and construction process, such as lack of coordination between disciplines and systems, overdesign of systems, incomplete design, or designing over the allowable budget.[22]

Critical components of TVD include:

- **Defining Client Value**—"Engage deeply with the client [...] Both designers and clients share the responsibility for revealing and refining concerns, for making new assessments of what is value, and for selecting how that value is produced."[23] This process includes recognizing and grappling with the client's business plan. "During business planning, owners define what they must have in their facilities and what they require for a return on investment (ROI) to establish their true allowable (i.e., bottom-line) cost for projects. They decide whether to proceed, and select a team to execute validation."[24]
- **Setting the Target Cost**—According to Umstot and Fauchier, a target cost is set by reducing the client's Allowable Cost by some percentage, perhaps as much as 15 to 20 percent as a stretch goal.[25] Sutter Health Care and DPR report a different and more intentional process, one called

validation. During validation, "the integrated team creates conceptual design alternatives and seeks to determine whether the project may be built for the allowable cost. It provides cost-saving strategies that may range from prefabrication and modularization to devising more efficient workflows to create the most value for the dollar. Typically, all of this is compiled into a report that can be presented to top management and a board of directors."[26]

All the literature reviewed on TVD concurred that once the project the Target Cost is set, the project estimate is then broken into building systems components. Each has its own estimate used as individual targets totaling to an amount equal to the Target Cost, which is below the Allowable Cost. According to Umstot and Fauchier, "The cardinal rule of target costing is that the target cost must never be exceeded."[27]

- **Rapid Continuous Estimating**—TVD depends on a fully integrated and collaborative approach to problem-solving and decision-making. The Big Room is a typical tool used by projects using TVD. Lean Project Consulting advises project stakeholders to "design to a detailed estimate. Use a mechanism for evaluating design against the budget and the target values of the client. Review how well you are achieving the targets in the midst of design."[28]

DPR reports that "Cost estimating and budgeting are continually updated, and every decision made must be measured against the cost target. Cross-functional teams comprising designers, builders and owner's representatives are put together, often referred to as 'clusters.' "[29]

Key advice for using TVD:

- The owner must be deeply involved and ready to make decisions that support the values, principles, and goals agreed to by the team.
- The owner should empower the team to make decisions that are in line with values, principles, and goals agreed to by the team wherever possible so that certain problems can be attacked on the sub-team level.
- Leaders should stay open to new ideas from the teams to add value. The ideation for cost savings and value creation does not stop at the validation phase.
- All the lessons learned in the Leadership and Collaboration section make TVD more powerful.

Decision Frames Overview

Whether you are using one of the decision frames discussed here or have selected another frame for your project, certain common principles are critical to success:

- The owner must be deeply involved in the decision process and understands the consequences of not making decisions.
- The decision process must be collaborative and inclusive.
- Once a decision frame and related methodology have been chosen, stick to it. Compromising will endanger the team's trust.
- Decision-Making is grounded in Decision-Ready information.
- All the lessons learned in the Leadership and Collaboration section make any decision frame more powerful.

Decision-Ready Information

Kevin Neumann was a college junior on the University of Arizona baseball team in the spring of 2015. In a few weeks, he would be drafted in the first round of the MLB draft by the Pittsburgh Pirates, and, in a few more years, he would realize his dream of playing in the major leagues for the Pirates. However, on this day in 2015, Kevin stood on third base in a tied ballgame with two outs in the bottom of the ninth inning against Rice University. The difference in the ballgame was ninety feet away at home plate. Runs had been hard to squeeze out in this game against Rice, and Kevin knew, as did his legendary coach, Andy Lopez, that waiting for a base hit to win the game might not be the best decision.

Coach Lopez needed more information before he could determine if there was another alternative as he surreptitiously drew a stopwatch from his pocket. He started the watch when the pitcher on the mound began his windup motion. The coach stopped the watch when the ball reached the plate. Glancing at the time, he reset the watch and timed the pitcher's motion for another pitch and then another. Confident he had enough data to make his decision, Andy Lopez gave Kevin the sign.

Kevin once again worked his way down the third-base line as he did before each pitch. Everyone expected Kevin would return to the base unless the batter hit the ball. This time, however, Kevin broke into a full run the moment the pitcher started his windup. Kevin stole home before the ball reached the catcher's mitt. Once Andy Lopez knew that Kevin could run fifty feet in less time than it took the pitcher to deliver the pitch, the decision was worth the risk. The stopwatch provided Andy with the needed decision-ready information.

What Do We Need?

In the design and construction industry, we often face an owner or team that will not make a decision. Many projects have failed not because of poor design or poor construction planning but because decisions are not made in a timely fashion. We have already reviewed the risk of not making decisions, how important it is to empower the entire team to make decisions, and the decision frames and principles that will enable those decisions. However, none of this advice will help the project team if the team does not produce decision-ready information for making decisions when needed.

I did not understand this concept until I attended business school to earn an MBA in the late 1990s. In architecture school and after, we continuously deal with critical design decisions that we believe are relevant to the client. However, I did not, in architecture school, get an understanding of the business needs of the client. Therefore, like most architects, I did not reflect, in design drawings or communications, those critical issues that would drive decisions for our clients. As Cramer and Simpson wrote in *The Next Architect*,

> there are many important aspects and implication of the design that are not evident anywhere in the drawings. For example, what will the building actually cost? How long will it take to build? How much energy will it consume? How much will it cost to be maintained over time?"[30]

As with decision values and principles, decision-ready information will not be the same for every client. There will always, however, be some consistencies, as Simpson stated in a recent interview. Decision-ready information must include

- two to three most viable options,
- impact on project budget for each option, and
- impact on schedule for each option.[31]

This outline is a starting point, but we should dig deeper. When you consider budget impact, all costs related to the proposed change must be considered, including design fees, rework, permit costs, and, as appropriate, contingencies. The owner mustn't be surprised later on with hidden costs or "forgotten" costs.

Decision-ready information must include functional and long-term financial impacts as well, which means considering the needs of all the stakeholders in the process and the needs of different owners. Some owners, like university owners, will need long-term maintenance implications, so that information must be captured. Sometimes, decisions have an impact on a business plan or direct revenues. Stadiums are a great example of this concern. Certain seats can draw more revenue than others. VIP boxes can draw licensing revenue. Points of sale for concessions are critical. Therefore, every design approval must be accompanied by a revenue plan.

Some owner programs will require flexibility but still have minimum performance requirements. Schools are often in this category. Therefore, decision-ready information will have to include the impact on those minimum requirements or on the facility's flexibility (e.g., multiuse gymnasium or performance space). In determining a floor-to-floor height or shaft space for laboratories, it may be critical to understand the cost versus future flexibility in types of research that can occur in the space.

Regardless of the project or the owner, it is advisable to ask all significant stakeholders what information is needed to make any major decisions. Anticipation and planning are vital. Do not assume you know what the owner or other stakeholder need.

Use your imagination and curiosity as well. Preparing decision-ready information requires a curious nature and a willingness to ask "why" five times. Like Coach Lopez, ask what critical piece of information will inform this decision and make the decision worth the risk.

Decision-Ready Information must include an assessment of the decision risk.

Decision Risk

Any time a decision is made, there is always a risk that the decision will be the wrong decision. It is possible to get many smart people in a room with excellent decision-ready information and still make an incorrect decision. Please note, I did not say it would be a bad decision because I would define a bad decision as one that was made without deliberation, adopted decision frames, or decision-ready information.

Great decisions are made using the guidelines discussed in this chapter, even when they are sometimes incorrect. I know that may seem nonsensical. However, moving decisions along in a project following these guidelines will be very powerful even when you make some mistakes. When mistakes are made, your team will have the tools and confidence needed to respond quickly and to make the decisions to correct those mistakes.

Regardless of the excellence of your team and your decision-making processes, you must recognize that there is some risk in making almost any decision. Often the risk is directly associated with a lack of information. You may take it for granted that no matter how dedicated and experienced your team is, there will be information that will not be available when you make a decision. Sometimes that information relates to future materials market performance. Sometimes it relates to a companion product not yet selected, or a system not yet specified or designed. Your team will face these challenges and, yet, have to make decisions without complete information.

Therefore, decision-ready information must include a decision risk assessment. Imagine a team is designing a medical office building for a developer, and the steel markets have been turbulent due to the international market. The risk of choosing a steel structure may be that although the steel price is very low during design, the price may be significantly higher when purchasing the structure. A concrete structure might be a few percentage points higher in cost at the time of the decision, but the price is stable. A stable price may have value to the owner. Discuss these kinds of risks as you make decisions.

Decision-ready information must also include the consequences or risk incurred if the decision is not made. For example, if the developer of the medical office building discussed above is hesitant to decide on the structure because she wants to watch the market for a few weeks, the developer must be informed of the risk of delay. The developer should know what parts of the design cannot move forward while the decision on the structure is delayed. Therefore, decision-ready information must include the risk of not making a decision at the scheduled time. Consequences must be described as precisely as possible. If your project loses $5,000 per day for a delay, state the per day cost and the total amount for the entire estimated delay.

The same principle applies to any change to a decision that has already been made. Certainly, there will be mistakes that must be fixed, but teams are often faced with the desire to revise a design decision. Changing decisions can be a bad habit for a team, but changes are not all bad. Therefore, when changes must be considered, decision-ready information must include the consequences and risks of changing the decision. Again, taking the example of the medical office building above, imagine the developer, after deciding on a concrete structure, saw the market for steel drop precipitously two months later. In this case, the decision-ready information must include not only the difference in cost of the material but all the costs related to rework of the design and the cost of the delay of the entire project. The risk of the market rebounding must also be considered.

Decision-Ready Information Must Haves:

- Two to three most viable options including pros and cons.
- Quantified impact on the project budget for each option.
- Quantified impact on schedule for each option.
- Functional impact to the owner's business of each option
- Financial impacts of each option, including all costs related to construction and ownership as well as the impact on the business plan.
- Decision risk related to making the decision and to delaying the decision

And for changing decisions:

- All costs related to changes including design fees, rework, permit costs, and as appropriate contingencies.
- Financial and functional impacts on the owner's business.
- Quantified impact on the schedule for making the change.
- Decision risk related to changing decisions

The Future of Decisions

The end of the college baseball season culminates with the College World Series in Omaha, Nebraska, one of the great traditions of college sports. So grounded in tradition is this pageant of America's favorite pastime that one might have imagined that technology might never invade those pristine fields. Yet, after some close calls by umpires in past years were perceived by some to have cost some worthy teams their championship rings, the "instant replay" arrived in Omaha to assist with some of the very close decisions.

Technology

Technology is changing our lives in design and construction decision-making in a surprising number of ways. Of course, Building Information

Modeling (BIM) has now made information sharing much more likely if not easier and it has provided us the ability to embed our models and drawings with data, lots of data to assist with decision-making on all levels. Pushing decisions to the ideal level and the ideal moment within each project becomes even more important when considering the opportunities presented by BIM and the convergence of technologies described by Randy Deutsch in *Convergence: The Redesign of Design*. Technologies and professional disciplines are converging and providing the opportunity to provide more predictable outcomes for projects. As Deutsch states, "With increasing demands to make decisions in real time, design professionals—having met the challenges and opportunities of this moment—are moving beyond the linearity metaphor and thinking in terms of simultaneity, superintegration, and convergence."[32] In this emerging version of the industry, it is critical that decisions are pushed throughout the project team to take full advantage of the power of convergence, asynchronous and deeply iterative design, and problem solving.

Deutsch points out that this convergence has its challenges.

> It requires the understanding of BIM as not only a document-generation tool but also, and more importantly, as a sharable database containing a project's information. Seeing BIM as the next generation of CAD keeps both practitioners and academics—the industry, profession, and education—from benefiting from the convergence. One of the factors distinguishing between CAD and BIM is that BIM is not only a technology, but also a process—a collaborative process [...] The BIM technology and process have led to innovative workflows that enable convergence to occur throughout the project lifecycle. Workflows that were decidedly linear have become iterative, and, as they make their way toward a solution, convergent.[33]

In *Managing Design*, John Moebes, senior construction director for Crate and Barrel, observed, in fact, that this convergence is hard to achieve.

> I'm seeing "BIM retrograde" these days. The millennial generation is drifting back to producing the same 2D documents we produced 15–20 years ago [...] BIM passion has subsided to a large degree. The adoption level is high, but the passion to solve problems has subsided.[34]

It is not surprising that the convergence of these technologies stalls within the design and construction industries if we have not pushed decision-making to the ideal levels of projects and empowered the team members to make decisions when they have the information needed. Decision flows are substantially linear, yet convergence allows for iteration and refinement. Teams must grapple with this tension and set rules for this engagement. Only then will teams be able to leverage these technologies to their capacities.

Data

The exponential growth of data availability is also a challenge for the industry. Many of our professionals are not trained to deal with or appreciate data and data analytics. Still, data alone do not always provide the answers we need in design and construction. In Scott Hartley's book, *The Fuzzie and the Techie: Why the Liberal Arts Will Rule the Digital World*, Hartley discusses the problems related to accepting the answers of algorithms, machine learning, and traditional analytics without a liberal arts or human-focused filter. Hartley announces, "As 'software eats the world,' technology requires input and expertise from every corner of society."[35] There is no doubt that as we move forward into the next decade, more and more of design will be informed by, if not produced by, artificial intelligence or machine learning. Randy Deutsch asks, "Where in the design process is human input needed and where is it redundant? What role will our legacy tools play and to what extent are they holding us back?"[36] I ask, how will these decisions be made in the future? It is clear that data-driven decisions must continue to be informed by human creativity and by human insights.

Both Hartley and Deutsch argue that algorithms are created by humans and are, therefore, subject to review and to challenge. This fact is not a reason for avoiding these decision tools but gives us a reason to embrace and inform them. Deutsch argues,

> What exactly is it that data does for architects and their building projects that standard knowledge, experience, and intuition can't? It eliminates paths that don't lead anywhere, and reveals hunches and assumptions that were enabled by preconceived notions that turn out to be incorrect. The beauty of gathering and leveraging data in building projects is that it enables designers to save time and valuable resources by eliminating false positives.[37]

Expanding our ability to use these tools to create decision-ready information will also improve our ability to serve our clients and our teams.

Dynamic Modeling

The year before I left my job at Rice University, we started on a unique decision journey to support the university in master planning. Working with KieranTimberlake, we took on the challenge of the complex future of the university. It was clear that a traditional master planning approach was no longer serving the university leadership as effectively as it should. The university's future would be a future of changing student and faculty ratios, pressurized parking needs, and increased stormwater challenges as well as a desire to increase the number of undergraduate and graduate students housed on campus. The university needed a tool to help decision-makers grapple with multiple scenarios for the future. KieranTimberlake designed a multivariable data-based dynamic model grounded in the university's values and operating principles that could be visualized in three dimensions. The Integrated Campus Plan, or ICP, presents itself to the user as a dashboard where individual scenarios can be shaped for a large number of variables. For example, one scenario may be to increase the percentage of students housed on campus but keep undergraduate enrollment level. Another scenario can reduce the percentage of cars permitted on campus by students and increase the area of green roofs on campus buildings. The model then calculates, assuming all other variables remain the same, the amount of new housing construction required, the total cost of construction, the impact on parking, estimated total energy consumption increase, and the impact on stormwater capacity.

The ICP is an example of a Dynamic Decision Model. Models work best when they are infused with the underlying values and "DNA" of the organization or stakeholder. Therefore, they are not easy to build. They are, however, worth the investment and will increasingly be used by planners and owners to make difficult decisions and to ensure a more resilient future.

PropTech

Property Technology (PropTech), the technology for real estate owners to support and streamline the purchasing, building, owning, and managing their portfolios, is a rapidly growing industry. This software is often built with the owner's needs in mind. If embraced by the project team, PropTech will support efficient decision-making. We can expect that PropTech developers will be among the first to leverage building data generated by the millions of sensors now being installed in our "smart" buildings. It's also clear that owners that use these new and emerging software resources will be requiring participation and buy-in by project team members to increase predictable outcomes.

Just-in-time knowledge delivery

Just-in-time knowledge, the notion of having the knowledge you need delivered when you are making the decision that requires that knowledge, is a critical need for decision-making in design and construction. We have already discussed that decisions pushed earlier in the project time line will provide the greatest value to the project; however, much of the information required to make those decisions, in traditional processes, is simply not available. Most pricing and product information, not to mention installation experience, is only available once the subcontractors come on board.

Collaborative project processes do a better job of delivering knowledge when needed, but this is still an imperfect solution. For example, a colleague recently raised the issue of the chemical incompatibility of newly developed materials. The time to understand if two materials are incompatible is when the details are designed and when they are specified.

Doctors receive current information about medications automatically when they diagnose and prescribe to their patients. Contemporary medical enterprise systems are designed to provide just-in-time knowledge to prevent misdiagnosis and inappropriate prescriptions. It would make sense to develop smart processes within BIM and specification programs to reach out and to provide just-in-time product knowledge for design and construction professionals when it is needed during design.

These are the kinds of challenges Katerra and other emerging technology-focused companies are figuring out—how to get genuinely useful and absolutely accurate information at the moment it is needed in design. Technology can help with this. Perhaps, this is the future of decision-ready information.

Summary

Master-level decision-making is a critically important component of the Culture of Predictable Outcomes. Without disciplined decision-making, leadership and collaboration have no purpose and are frustrated. What makes this decision-making master-level quality? It is practiced, exacting, and well-informed. Like a master chef, a master-level decision-maker is excellent at his/her mission. Like a master chef, a master-level decision-maker knows how to make something amazing out of almost any ingredients.

Great decision-making moves projects forward in a practiced, exacting, and well-informed manner. This decision-making discipline is about excellence, and, even when things go wrong, master-level decision-making can help you keep your project on track. Decision-making supporting a Culture of Predictable Outcomes is timely, well informed, values based and disciplined.

Postscript—Driving Decisions When You're Not the Boss!

What do you do when you are not the owner or a senior leader but want to influence the process? What if you are a subcontractor that wants to make a difference? Perhaps this advice will help.

Suppose you have been selected for an amazing design-build project. You are excited about the team. Both the contractor and the architect are known for collaborative practices, and the owner is known to be a good and thoughtful leader on projects. This should be a positive experience if only everyone will make decisions when they are supposed to. Nothing can be more powerful than a project where decisions are made quickly and effectively, yet we have all seen great projects fail because decisions were mired in the quicksand of debate, insufficient information, or, simply, lack of will. Nothing can be more frustrating. What can a subcontractor do to change or even prevent decision constipation occurring? Here are seven decision-planning strategies you can use even if you don't have a great team or a great leader with whom to work:

1 Earlier the better—Make every effort to discuss decision-making at the start of the project. Creating decision norms, frames, and tools will serve every team member and will prevent misunderstandings.
2 Yours, mine, and ours—Make it clear to your team what decisions you believe are your decisions to make, what decisions aren't yours to make, and how you want to handle decisions that have multiple stakeholders. Try to push decisions to the ideal, most effective location in every team. Discuss strategies to make sure that decisions quickly flow even if you are not sure who owns the decision.

3 Lean In—The best HyperTrack© (not Fast-track) decision processes encourage teams to schedule all significant decisions at the beginning of a project and to schedule who will need to be included in those decisions. This process is like creating a Lean Pull Schedule for decisions. Even if your entire design-build team does not buy into this strategy, you can lead with your decision pull schedule. This way, you have informed the whole design-build team when you will be making crucial decisions to keep the project on track.

4 Be decision-ready—Once you know when decisions will be made, even if it is just for your own work, you can plan to prepare decision-ready information. This often means informing the rest of the team of the information they need to prepare so that your decisions can be made. Recognize that decision-ready information means that you inform the contractor and the owner of all costs related to a decision when that decision is being made. Revealing a hidden cost later does not help the team, and it does not help you.

5 Ask…don't tell—When you aren't in a leadership position on a team, it is sometimes tricky to raise new ideas. The best way to bring others around to your way of thinking is to ask thoughtful questions about when and how decisions will be made on a project.

6 Explain the downside—The best value you can provide when a client or team is tempted to delay a decision is to explain the consequences of delay in that moment. The consequences must be tangible and quantifiable. Merely saying it will cost more won't cut it. Stating that a delayed decision costs $1,000 per day will get an owner's attention.

7. Stay on it—The most successful teams recognize that projects can go wrong in a moment. When surprises occur, find ways to help your team make corrective decisions quickly. Setting a team norm that states that you will never leave a meeting without getting the project back in budget and on schedule helps with that discipline.

Remember, decision-making is an essential element to success on any project. It requires discipline and focus. Even though the owner and the contractor will be responsible for most decisions, you can help lead your team to better, more successful decision-making practices.

Notes

1 Amy C. Edmondson and Salter Reynolds, *Building the Future: Big Teaming for Audacious Innovation* (Oakland, CA: BK/Berrett-Koehler, 2016), 107.

2 Barbara White Bryson and Canan Yetmen, *The Owner's Dilemma: Driving Success and Innovation in the Design and Construction Industry* (Atlanta: Ostberg Library of Design Management, Greenway Communications, 2010), 104.

3 Michael Alan LeFevre, *Managing Design: Conversations, Project Controls, and Best Practices for Design and Construction Projects* (Hoboken, NJ: Wiley, 2019), 31.

4 Barbara White Bryson and Canan Yetmen, *The Owner's Dilemma: Driving Success and Innovation in the Design and Construction Industry* (Atlanta: Ostberg Library of Design Management, Greenway Communications, 2010), 105–107.

5 "Crowdsource," Dictionary.com, www.dictionary.com/browse/crowdsource?s=t.

6 Barbara White Bryson and Canan Yetmen, *The Owner's Dilemma: Driving Success and Innovation in the Design and Construction Industry* (Atlanta: Ostberg Library of Design Management, Greenway Communications, 2010), 62–63.

7 Pauline Nee, "Cost Tracking Checklist for Real Estate Owners," Honest Buildings, https://blog.honestbuildings.com/cost-tracking-checklist-for-real-estate-owners.

8 Barbara White Bryson and Canan Yetmen, *The Owner's Dilemma: Driving Success and Innovation in the Design and Construction Industry* (Atlanta: Ostberg Library of Design Management, Greenway Communications, 2010), 157–165.

9 Ibid., 155–156.

10 Ibid.

11 David Umstot and Dan Fauchier, *Lean Project Delivery: Building Championship Project Teams* (Anacortes, VA: Armchair Publishing, 2017), 153.

12 Ibid.

13 Faaiza Rashid and Amy C. Edmondson, "Risky Trust: How Multi-Entity Teams Develop Trust in a High Risk Endeavor," Harvard Business School Working Paper, 2011

14 James P. Cramer and Scott Simpson, *The Next Architect: A New Twist on the Future of Design* (Norcross, GA: Ostberg Library of Design Management, 2007), 75.

15 James P. Cramer and Scott Simpson, *The Next Architect: A New Twist on the Future of Design* (Norcross, GA: Ostberg Library of Design Management, 2007), 75.

16 Scott Simpson, interview by telephone, June 17, 2019

17 "Target Value Design Gaining Ground," *DPR Review*, Fall/Winter 2012, www.dpr.com/media/review/fall-winter-2012/target-value-design-gaining-ground.

18 Ibid.

19 Martin Fischer, Howard Ashcraft, Dean Reed, and Atul Khanzode, *Integrating Project Delivery* (Hoboken, NJ: Wiley, 2017), 25.

20 David Umstot and Dan Fauchier, *Lean Project Delivery: Building Championship Project Teams* (Anacortes, VA: Armchair Publishing, 2017), 153.

21 Ibid., 154.

22 Ibid., 154–157.

23 Hal Macomber and John Barberio," Target-Value Design: Nine Foundational Practices for Delivering Surprising Client Value," Lean Project Consulting, 2007, www.leanproject.com.

24 "Target Value Design Gaining Ground," *DPR Review*, Fall/Winter 2012, www.dpr.com/media/review/fall-winter-2012/target-value-design-gaining-ground.

25 David Umstot and Dan Fauchier, *Lean Project Delivery: Building Championship Project Teams* (Anacortes, VA: Armchair Publishing, 2017), 157.

26 "Target Value Design Gaining Ground," *DPR Review*, Fall/Winter 2012, www.dpr.com/media/review/fall-winter-2012/target-value-design-gaining-ground.

27 David Umstot and Dan Fauchier, *Lean Project Delivery: Building Championship Project Teams* (Anacortes, VA: Armchair Publishing, 2017), 157.

28 Hal Macomber and John Barberio," Target-Value Design: Nine Foundational Practices for Delivering Surprising Client Value," Lean Project Consulting, 2007, www.leanproject.com.

29 "Target Value Design Gaining Ground," *DPR Review*, Fall/Winter 2012, www.dpr.com/media/review/fall-winter-2012/target-value-design-gaining-ground.

30 James P. Cramer and Scott Simpson, *The Next Architect: A New Twist on the Future of Design* (Norcross, GA: Ostberg Library of Design Management, 2007), 81.

31 Scott Simpson, interview by telephone, June 17, 2019

32 Randy Deutsch, *Convergence: The Redesign of Design* (West Sussex, UK: Wiley, 2017, Kindle ed.), 458.

33 Ibid., 2326–2337.

34 Michael Alan LeFevre, *Managing Design: Conversations, Project Controls, and Best Practices for Design and Construction Projects* (Hoboken, NJ: Wiley, 2019), 26.

35 Scott Hartley, *The Fuzzy and the Techie: Why the Liberal Arts Will Rule the Digital World* (Boston: Mariner Books, 2018), 58.

36 Randy Deutsch, *Convergence: The Redesign of Design* (West Sussex, UK: Wiley, 2017, Kindle ed.), 551.

37 Ibid., 1183–1190.

Chapter Supplement—4
On the Right Track

In 1999, Scott Simpson, CEO of The Stubbins Associates, was enjoying his second cup of coffee of the morning and lingering just a few extra moments over his newspaper, the *Cambridge Chronicle*, when an article about local real estate caught his eye. He read the story a second time and then picked up the phone to call an old high school classmate for what was to be a $30 million conversation.

The article outlined a plan to down-zone a portion of Cambridge near Kendall Square. The proposal would adjust the current 4.0 FAR to a much more modest 2.5 FAR, essentially devaluing the affected properties by more than a third. Scott knew his high school friend was CEO of Amgen Inc., a company that owned property in the impacted area. So, when he reached Kevin Sharer, the Amgen leader, on the phone, Scott explained how FAR changes could result in significant asset losses. Kevin responded quickly, sending Amgen's development manager, Pete McCawley, to Cambridge. Pete had one question, "How can we save the development potential of the property Amgen owns?"

Scott's phone call was fortuitous in many ways. According to the *New York Times* in February of 2000, the Kendall Square area was soon to be a hotbed of research space construction. Pharma companies seemed happy to build and own facilities in the area at approximately $350 per square foot or rent from others at a new high of $50 per square foot (prices are significantly higher in 2020). As Susan Diesenhouse, writer for the *New York Times*, noted in an article in 2000, "The lure for this new construction in Cambridge is the intellectual firepower of the Cambridge workforce, forged at some of the world's top colleges, research institutes, medical schools, and teaching hospitals here and across the Charles River in Boston."[1]

Pete McCawley was quoted at the time as saying, "To keep our pipe-line full at Amgen, we want to be in the Cambridge market [...] it's no secret that Cambridge has the largest and best pool of scientific research talent in the world."[2] The same article went on to state that vacancy rates in Cambridge labs were extraordinarily low (1 percent) in late 1999. The hot market increased the urgency for Amgen to maximize the development of its property.

The team only had thirty days to save the development potential of the property. The City of Cambridge had reacted to the biotech boom by moving to prevent overbuilding. The first step would be an eighteen-month moratorium on new permits for projects larger than 20,000 square feet in the Kendall Square area. The second step would be to pass a new zoning code with the reduced FAR. Before the FAR reduction, the Amgen property would accommodate nearly 300,000 square feet of space. The new restrictions would reduce that by at least a third, equating to a loss of approximately $30 million in asset value. To prevent this from affecting their development, Amgen would have to submit construction documents for a permit within thirty days, a seemingly impossible task.

Scott Simpson, as he recollects those next thirty days, gives Pete McCawley credit for sharing in and inspiring the development of HyperTrack©, the copyrighted process of decision-making that made the success of the Amgen project possible. Scott also gives Pete credit for being present every step of the way over the next month, making decisions and providing leadership as required.

As the reality of the thirty-day schedule hit home, planning started with purpose. As Scott remembers,

> Pete and I went out for a beer and did some thinking. We said, "Okay, let's assume that we're going to make this deadline. What has to happen, working backward from the end of February to the end of January, so we can file for the building permit?"

During that conversation, the two experienced leaders deconstructed the design and documentation process, breaking it into pieces. They soon realized the key ingredient to speed would be making decisions at the moment they were needed. Scott and Pete recognized they did not have the time for the typical iterative design process of developing multiple alternatives and testing back and forth for the preferred solution. That was a luxury they did not have.

So, Pete and Scott designed an alternative process. The City of Cambridge would only require construction documents for the shell and core of the building within the thirty-day window. The limited scope allowed the effort to be divided into two parts and two teams. The teams were nicknamed the "bullet train" and the "local train." The "bullet train" team worked on the shell and core and had to submit drawings in thirty days without fail. The "local train" team worked on Tenant Improvement (TI) drawings and could lag a bit as user groups provided additional information.

To support this effort, Pete McCawley moved to Cambridge for the duration of the project. Scott's firm found him an apartment just two doors down from the Stubbins office. They also hired a secretary that followed Pete around all day every day. The Stubbins team needed 24-7 access to Pete, which meant, in 1999, phone and fax. Stubbins also provided Pete a desk in their office. Pete and Scott also hired a lawyer, an expert in zoning, to attend every project meeting and to make sure the team was appropriately informed of the zoning implications of every design decision. Finally, Scott and Pete hired a contractor who attended every meeting and participated in all design decisions.

Access to decision-makers, stakeholders, and special consultants was just the start of the new HyperTrack© process. The new process also required new discipline. Scott and Pete established "clear and consistent decision-making protocols" for the team requiring that every meeting have a preset agenda outlining every key decision to be made during the meeting. Decision-makers or their designated representatives were required to attend the meetings. If they could not, the meetings were cancelled. All key stakeholders attended meetings. Decision-Ready Information, pros and cons, as well as impact on cost, schedule, and quality were required for every decision. If decision-ready information was not prepared, the meeting was cancelled. Simpson noted, "Once a decision is made, the [...] team is authorized to take immediate action to implement it."

The essential ingredients of a good research building were not hard to sort out. The basic floor plan and floor-to-floor heights were the first items to be designed. Then the teams took on the basic circulation system including the elevators, stairs, and corridors. "These were the bare bones of the building," Simpson observed later. Once those aspects were determined, the team examined the facade. Taking on these challenges, the team worked quickly and carefully, devising meeting agendas to inform all team members of the critical issues that were being investigated and which decisions were being made. "For example," Scott said,

> Monday, it might be the site coverage. The goal of that meeting would be to determine the footprint of the building. The second day, it would be the floor to floor height. The third might be the basic circulation system, and we would determine that.

This unique design approach and decision process would be tested early on during the Amgen project. The pharmaceutical company had not yet decided whether the building was to be an office, a laboratory, or a combination of both. Separating the shell from the TI, the "bullet train" from the "local train," helped the team stay focused. The team did not need to know where every fume hood or every office would be placed to approve a shell and core floor plan. The "bullet train" team, to a large extent, designed as if the tenant would come along later, similar to a spec office building. However, significant decisions that would accommodate the systems of both labs and offices had to be considered. Because the floor-to-floor heights were designed for lab space, if the building eventually became an office building, it would have to absorb the large duct spaces required by a traditional mechanical system with penthouse air handlers. That seemed unacceptable. Instead, the architects developed a "zebra system" where the building was treated as a layer cake with two mechanical rooms at either end that distributed air horizontally, and exhausted vertically for each floor. According to Scott Simpson, this design freed up a "great deal of usable square footage inside the building for either lab purposes or for office purposes." Strategies like this helped the team

> get a BOMA rating on the Amgen lab of 85 [percent]—a remarkable feat when most labs scored 72 [percent] or less. That additional BOMA square footage was a huge financial gain for the client and really raised the value of the building.

Having a clear path for decision-making was an enormous help. The team worked every single day and sometimes far into the night. Interestingly, Scott reflects that it was a great deal of fun. There were many team dinners and outings to sports events. The rules of engagement for meetings were clear. Everyone would be prepared, and nobody would be late. If you were late, the consequences were that you had to sing a song, often resulting in hilarious outcomes and additional bonding.

The team was also well motivated. The reality was that the building itself was a political target. Neighbors of Kendal Square were justifiably worried about the area being overbuilt. Community groups were dead set against a building being built on the Amgen site and there was the demolition of an historically significant building involved. Many community meetings were held during the thirty days the team was racing toward their deadline. Pete McCawley took the lead in these communications. He wanted both the community and the project team to understand that Amgen and the building had a purpose. That purpose was defeating cancer. Early on, Pete coined the phrase "the patients are waiting." The building would help Amgen do its best to deliver life-saving cancer drugs to those who need them. Pete reinforced that the sooner Amgen could get into the building, the sooner the scientists could get working, and the sooner the medicines would be developed and delivered. Pete's leadership and vision were effective at both the team and the community level, motivating both to support and deliver the building.

When the thirty-day deadline arrived, the team delivered the drawings for the shell and core to the City of Cambridge. Amgen received a permit that preserved the asset value of the property. The Stubbins Associates' first experience with HyperTrack© was a remarkable success and it would not be their last.

Notes

1 Susan Diesenhouse, "Space for Research Is Expanding in Cambridge," *New York Times*, February 13, 2000, www.nytimes.com/2000/02/13/realestate/space-for-research-is-expanding-in-cambridge.html.
2 Ibid.

Aligning Values and Goals **5**

When I started writing Creating a Culture of Predictable Outcomes, there was no Astros sign-stealing scandal, there was no COVID- 19, and, in the summer of 2019, George Floyd was still walking in the world. At that time, I was focused on the three foundational elements of a *Culture of Predictable Outcomes*, leadership, collaboration, and decision-making. Although I understood the importance of values, I had planned, at that moment in time, to integrate values into the leadership or collaboration chapters of the book instead of dedicating a separate chapter to values as Canan Yetmen and I had done in *The Owner's Dilemma*.

But the recent events in the world have given me pause. It all started with a comparatively unimportant event. When the news of the Astros' sign stealing scandal came out in January 2020, I had to stop and wonder. How had an organization that seemingly had it all, strong leadership, an amazing even charismatic team, and analytics-based decision-making, screwed it up so badly? Then I encountered an article in *The Athletic*, from October 2019, titled "Taubman Saga Exposes Longstanding Questions about the Astros' Culture under Jim Crane and Jeff Luhnow." This article acknowledged the strengths of the organization but noted the "winning at all costs" mentality. The author wondered aloud, "What was to prevent someone from pushing the wrong boundaries?"[1]

The Astros' values, the wrong values, had been a problem in the organization for a while, ultimately materializing in the trash-can-lid, sign-stealing scandal. What were the costs of the wrong values to the Astros organization and team members? Along with losing face and credibility, there were suspensions and job losses at the highest levels, even after players and managers had moved on to other teams. There will also be financial losses in attendance and advertising revenue. When the team won the World Series in 2017, it appeared the organization had everything going for it, everything except well-placed values.

I highlighted several times in previous chapters the challenges we face in the design and construction industries, precisely because of industry fragmentation and our susceptibility to economic downturns. As COVID-19 shut down the economy in 2020, our industries were again scrambling to cope. Hiring was stopped, and we were suddenly looking for ways to add value as real estate prices and work environments dove into uncertainty. Soon, we imagined, we would be working with new teams and new clients in unfamiliar ways. The demands and stresses might be very different and the working conditions downright peculiar. Off balance and uncentered, we need sophisticated-caring leadership, high-performing collaboration, and master-level decision-making in the coming months and years. But most of all, we need to wrap all these elements in clear, strong, and loudly stated values. Why are we doing this? Who are we benefiting? What are we willing to give up? What are we not ready to give up? Have you aligned your values to your goals? Values can help us navigate our future. As Roy E. Disney, cofounder of The Walt Disney Company, so accurately stated, "When your values are clear to you, making decisions becomes easier."[2]

When Don Shula passed away in May of 2020, it was Shula's values Larry Csonka, the great Dolphins' fullback remembered,

> He [Don Shula] was always looking for what he called "the winning edge," but he had a sense of integrity to go with it. He wanted to win, but only if it was within the rules. He wouldn't have any part of bending any of the rules, let alone breaking them. That's who he was—winning meant everything, but not at a cost to his principles. That's why he meant so much to us.[3]

The winningest football coach of all time would not bend the rules. Shula's values and his goals might have seemed at odds to some. On Shula's team, his values and goals were perfectly aligned and clear to all his players.

In *The Owner's Dilemma*, I recalled a conversation that spurred me to think about values. I wrote,

> One question […] was asked quietly after the formal [Design Intelligence] meeting had broken up. Charles Dalluge [now president and COO of DLR] approached me with a thoughtful look and said, "I am interested in what you do, but I am even more interested in

how you do it. I mean, how do you create the environment where these actions and processes are possible?" That question is, as they say, the elephant in the room. No innovation is possible without a values-driven environment. Presidents of successful corporations and institutions understand the importance of values in an organization, but few project teams are led with values in the forefront.[4]

I still believe these words to be true.

As I noted in that book, a great deal has been written about values in business and service environments. It is well-traveled ground by many business scholars. Among the most important concepts, however, is that leaders cannot be everywhere. You cannot possibly create enough rules or give enough orders to cover every problem, every decision, every challenge, or every opportunity your team will encounter. No one can anticipate every potential circumstance a team member will face. As a leader and a collaborative team member, you will have to trust your fellow team members to use their best judgment. The only glue you have is a set of values to guide your team members as they handle the ambiguous situations they will undoubtedly face.

In the Astros organization, the owner, Jim Crane, fired both the general manager and the manager not for implementing the sign-stealing scheme that scandalized major league baseball but because "neither of them did anything about it."[5] It was clear from the MLB commissioner's report that the activities of the bench coach and others on the team had occurred with passive approval. The players had gotten the message that winning at all costs was their mission, even if it had not been clearly spoken. There were even moments when the manager, Hinch, ineffectually attempted to damage the monitor the players and bench coach were using to cheat. Yet Hinch never directly ordered the team to stop.[6]

Unspoken messages about values undermine us every day in the design and construction industries. When teams and leaders do not expressly, explicitly discuss values, assumptions are made, based on actions, of owners, project managers, even other team members that may go uncorrected. In *Dare to Lead*, Brené Brown warns, when "organizational values are gauzy and assessed in terms of aspirations rather than actual behaviors that can be taught, measured, and evaluated," we are practicing behavior that will get in our way of achieving our goals.[7]

Thomsen and Sanders note in *Program Management 2.0*, without clearly defined values,

A project or program team will be driven by both the stated and unstated values of the program's leaders. And it's likely that the owner has chosen leaders that reflect the owner's values. However, these values will differ from one leader to the next. And they will change if leadership changes—either in the owner's organization or the management team.[8]

This shifting or lack of clarity of values can be devastating for a team. It can undermine hard-won psychological safety and trust.

Clearly stating values is a challenge few organizations do well, as Thomsen and Sanders point out,

> often the mission statements are motherhood and apple pie, the policy manuals provide little insight into values and the management plan is incomplete or dated—providing flawed guidance [...] More often, values are expressed informally, and they affect behavior through intangible feelings and traditions. They are expressed through nuances in decision-making or comments in coordination meetings. They reveal subtle attitudes about risk, innovation, responsibility, work ethics and interpersonal relationships. Most will be unwritten and must be inferred from day-to-day interactions.[9]

Experienced professionals understand that values manifest in very specific ways and may even be conflicting. Higher education clients may need to save money, yet they want the best for their researchers and students. Hospitals will want to prioritize the patient experience and safety, yet they need to attract the best doctors by diverting resources to laboratories and offices. All of the goals, even the conflicting goals, may be worthy, but team members need to understand the underlying values that will drive decisions and guide behavior throughout the project. Will the owner save money even at the expense of the contractor's well-earned profit? Will the architect fight for a design element that does not align with project goals?

Spending time diving into values and understanding priorities, potential conflicts, and how to sort out ambiguous areas is critical to a team's success. When someone says, we want to win at all costs, do they mean at the cost of all of our futures? You want and need team members to be able to frame these questions for themselves. Their response should be clearly understood and shared values.

To avoid values confusion, follow these steps:

- discuss values early
- sort out and understand priorities

- anticipate potential conflicts
- communicate values consistently and continuously

Because I believe these are worth repeating, I provide my list of standard project values from *The Owner's Dilemma* for your consideration. A detailed discussion of each is provided in that book.

Strive for transparency because information is power.
Tell the truth … always!

Lead with exuberance and caring consideration.
Manage clearly, consistently, and cooperatively.

Create a safe, it's-OK-to-screw-up environment.
Attack processes, not people.
Balance risk and reward.
Take responsibility: Put the target on your own forehead.

Seek excellence.
It's not just about the building.
Plan for excellence.

Make decisions at the most powerful (efficient!) level and the most powerful moment.
Encourage a culture of curiosity and knowledge.
Collaboration means everyone.

Be tough when it is important to the team's success.
Defend the pack.
Have difficult discussions early.
Don't go to bed mad.

Build and preserve relationships.
Your mom was right—thank-you notes are important.
Share a compliment. What does it cost?
Celebrate milestones. Give people what they need to do their jobs.
Every discussion should end with the question *What do you need from me?*
We don't need a hero (*we need to know if you are in trouble*).[10]

I suggest each team work on their own list with their team and their project in mind. Is communicating to the stakeholders on a regular basis important? Is there a priority regarding the climate?

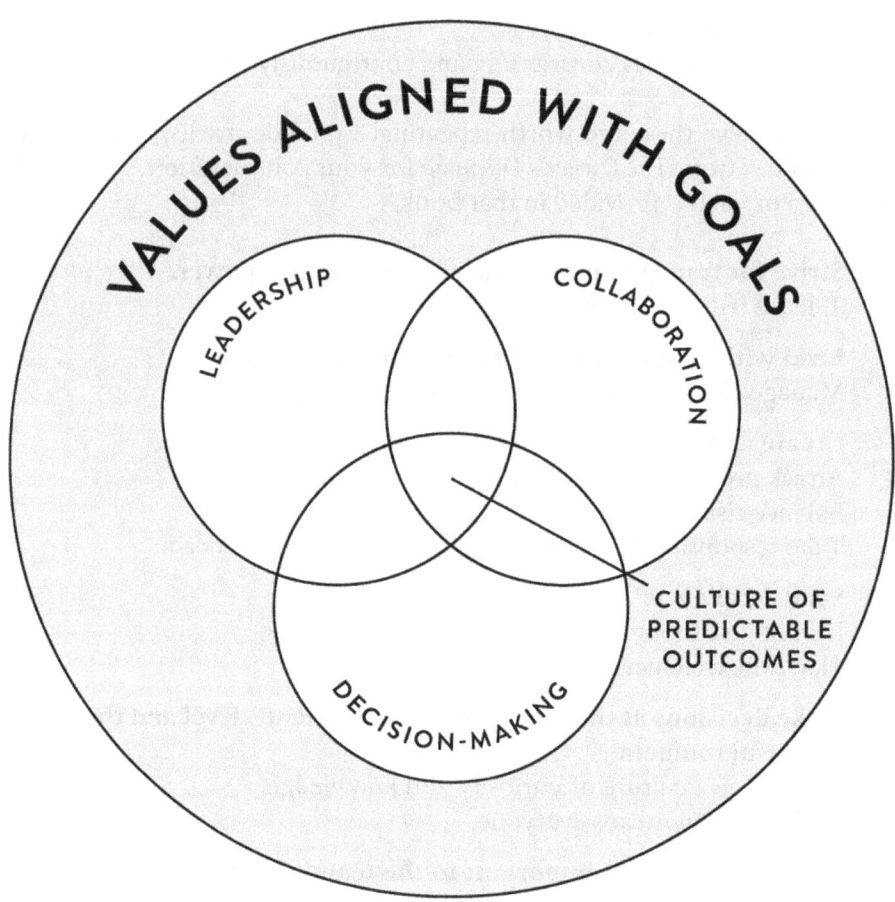

TO AVOID VALUES CONFUSION,
FOLLOW THE STEPS:
1. Discuss values early
2. Sort out and understand priorities
3. Anticipate potential conflicts
4. Communicate values consistently and
   continuously

Figure 5.1 Steps to avoid values confusion

Given the events of the last few months, I also suggest considering the following:

Strive for and value diversity in every aspect of our team
Work with people that don't look like you or think like you
Be kind
Listen

Play the long game
Think about the long-term consequences of your actions
Think about potential unintended consequences

Wear a mask (during a pandemic) because it protects other people
Try to stay team-focused rather than self-focused
Take action to benefit the team even if it makes you feel uncomfortable

In the design and construction industries where we are continuously reengaging with new teams and new people, we cannot expect anyone to read our minds and expect that they know our values unless we state them. Although I have always believed in the aphorism "Never wear your emotions on your sleeve," paradoxically, we absolutely must display, articulate, and repeat our values at every opportunity to make sure we invite diversity, maintain psychologically safe environments, support and protect healthy debate, and help team members function—even when they are off working in ambiguous environments facing new challenges.

Notes

1 Evan Drellich, "Taubman Saga Exposes Longstanding Questions about the Astros' Culture under Jim Crane and Jeff Luhnow," *The Athletic*, October 25, 2019, https://theathletic.com/1317907/2019/10/25/taubman-saga-exposes-lon…questions-about-the-astros-culture-under-jim-crane-and-jeff-luhnow/.

2 Joseph Demakis, *The Ultimate Book of Quotations* (Charleston, SC: CreateSpace, 2012), 276.

3 "Dolphins, NFL Family Pay Tribute to Don Shula," Miami Dolphins, www.miamidolphins.com/news/don-shula-remembered-by-players-nfl-executives.

4 Barbara White Bryson and Canan Yetmen, *The Owner's Dilemma: Driving Success and Innovation in the Design and Construction Industry* (Atlanta: Ostberg Library of Design Management, Greenway Communications, 2010), 117.

5 Jenna West, "Astros Fire A.J. Hinch and Jeff Luhnow Following Suspensions for Sign Stealing," *Sports Illustrated*, January 12, 2020, www.si.com/mlb/2020/01/13/aj-hinch-jeff-luhnow-fired-astros-sign-stealing.

6 Evan Drelich, "Hinch, Luhnow Fired as MLB Report Reveals Details of Astros Sign-Stealing", *The Athletic*, January 13, 2020, https://theathletic.com/1531133/2020/01/13/hinch-luhnow-fired-as-mlb-report-reveals-details-of-astros-sign-stealing/.

7 Brené Brown, *Dare to Lead: Brave Work. Tough Conversations. Whole Hearts.* (New York: Penguin Random House, 2018), 7–9.

8 Chuck Thomsen and Sid Sanders, *Program Management 2.0: Concepts and Strategies for Managing Building Programs* (rev. ed.) (McLean, VA: CMAA, 2011), 551.

9 Ibid., 554–555.

10 Barbara White Bryson and Canan Yetmen, *The Owner's Dilemma: Driving Success and Innovation in the Design and Construction Industry* (Atlanta: Ostberg Library of Design Management, Greenway Communications, 2010), 118–119.

Chapter Supplement—5
Conflicting Values

Note: The following story is inspired by actual events but, in truth, combines two separate but similar stories on different campuses. Therefore, the names and facts of this narrative have been changed to protect the institutions and the participants.

Shawn Gibson looked long and hard at the elegant invitation in his hand. Beautifully printed and crafted, an overlay ingeniously slid across color and pattern transitioning the design from dull to shiny, from delicate to bold as it was viewed. However, Shawn did not appreciate the cunning appeal. To him, this summons was a dark reminder of the disappointing events of the last two years.

The small Midwestern college was recognized throughout the world for its beautiful campus. Students and faculty were often attracted to the bucolic Edward Olmstead-designed campus the moment they stepped within its historic hedges. The collegiate Gothic architecture that dominated the campus sat gracefully among the sprawling lawns, towering oaks, luxurious planting beds, and flowering shrubbery. This campus had a long, precious architectural history, and Shawn Gibson, the college architect and facilities director, understood the value of this history. He also appreciated the charge he had received from the chairman of the board of trustees when he had been hired a dozen years ago, "This campus is part of our endowment. It constitutes a competitive advantage for the College. Don't screw it up."

To Shawn, that meant instilling specific values throughout his team. The College had a mission, and that mission of teaching and research came first, but for the facilities and construction team, providing excellence within the built environment that respected and leveraged the campus history was a close second. To Shawn, this did not mean the campus should be treated as a historical site, untouchable and unchangeable. Instead, it meant changes and additions should be made thoughtfully, maximizing what was already in place and working synergistically with the existing embedded architectural principles, all in a cost-effective manner. To Shawn, this directive also meant keeping the College's options open, keeping the institution flexible to cope with the challenges of the future. To some, a hundred-year-old building on a college campus represented permanence. To Shawn, the hundred-year-old building was an opportunity to create flexibility for the College's future, accommodating program needs and budget needs.

Shawn had other values as well for his organization and team. These values were about diversity, inclusiveness, transparency, listening, and caring. He fiercely defended his team but personally took responsibility for the team's mistakes, although that didn't happen very often. Shawn's teams believed in the value of excellence and in continuously improving their work.

For more than a decade, Shawn had used these values to keep all the College's competing priorities in balance, delivering more than twenty major new buildings for the campus and many more renovation projects. Shawn communicated these values early and often to every project team long before the space programming process started. Shawn also reflected these values to the senior administration and the board regularly, just to make sure everyone was comfortable with his approach. Shawn was willing to listen to other ideas, but so far, no one had offered a competing set of values to those his teams were embracing daily—until this project.

A Nasty Surprise

Shawn was suddenly confused and very concerned. He saw a storm brewing on the horizon and wasn't sure how to avoid it. In fact, from the earliest discussions on this new project, he had a terrible feeling this project would challenge all of his values. The project was, simply expressed, a piece of art, but it was also a building. Two members of the board of trustees, art patrons, were advocating for the College to build a unique art installation on the campus. The artist for the installation was world famous and known for her eccentric personal style. She had become friends with both of the wealthy board members who were in a position to donate significant funds toward an installation. Her work integrated technology and framed views to nature-like plantings, which were experienced from the inside of the installation. The technology provided specific vibrations and pulses in concert with music and small changes in the views. Each installation was a ballet of moving parts, pushing the capabilities of existing technology and architectural craftsmanship. It was all exposed to the elements and would be a nightmare to maintain.

The first surprise Shawn experienced was the site that was selected for the project, without his participation. Shawn was in charge of the college master plan, and under his direction, every building site was anticipated well in advance. As the college campus expanded, the original ideas and concepts from the Olmstead designs inspired planners developing new designs. Few decisions were made without long conversations with the board and close examination of future outcomes and impacts. So, when Shawn encountered Maryanne Jackson, the director of campus fundraising, turning circles with building plans in the middle of the lower gardens on Friday afternoon, he was a bit startled. Maryanne, it seemed, was waiting for the two board members and the artist to confirm the location for the new installation, which was to be in the center of the lower gardens.

Shawn was shocked. Not only was he surprised to find out that the art installation was a serious and significant proposal for the College, but he was also deeply concerned about the selected site. The location was directly in the middle of one of the loveliest areas of campus, also slated to be developed as a new research zone. Buildings had not yet been designed, but the intention was to work with the layout of the lower gardens, which were designed to be synergistic with the Olmstead plan. Shawn couldn't think of a worse place to locate the contemporary art installation, which would ruin the gardens as well as limit the development of the research area due to the view corridors required to make the art piece work. As Shawn began to explain these concerns to Maryanne, she curtly informed him that he was not invited to the meeting. She waved him away saying she would call him later.

As Shawn retreated, his head was spinning, but his hand was pulling his mobile phone from his pocket. He texted his boss, the vice president of finance and administration, wondering if he knew what was going on and, if he did, why hadn't he given Shawn some sort of warning. This new project, this art installation was progressing outside the typical board of trustees process, yet two board members were involved. Shawn began to assess the benefits and problems related to building an installation by this artist.

The Pros and Cons

There was no doubt the artist was known worldwide. Her pieces were widely published, and a couple had already been installed on other university campuses. An installation by this artist would draw international attention and a significant number of visitors to the College. This installation would also advance the vision of a few board members to have art play a more significant role on the campus, a worthy goal in Shawn's mind. He had participated enthusiastically in developing the public art program, identifying sites, and overseeing the installation process of new sculptures and two new fountains designed by artists. He also worked with the public art director to develop maintenance budgets for outside art, which had no funding source except his budget. Right now, the art pieces already placed on campus competed with building and landscape maintenance needs when they needed refurbishment.

This project was another matter altogether. An art installation such as the artist envisioned would be a permanent piece, actually a building. Unless the artist agreed to sign a contract that allowed the college to relocate the installation in the future, which was unlikely, the installation would be sitting in the lower gardens forever. The piece would permanently inhibit development, and it would forever require maintenance, including maintenance of the technology, which would inevitably become outdated. There was a good chance the completed installation, which could cost as much as $5 or $6 million, not including long-term maintenance and operation costs of thousands of dollars per year, would divert precious resources away from students, teaching, and research unless fundraising also included an endowment.

Shawn wondered if the president, vice president, general counsel, and the rest of the board understood all the issues. Shawn looked down at his phone as he walked toward his office on the edge of campus. "Don't worry. Got it!" was the only response from his boss. Shawn wondered what the heck that meant.

As the weeks passed, Shawn wondered if he had only imagined the impulsive movement on the art installation project. Everyone had gone dark on the subject. Maryanne wasn't answering Shawn's e-mails, and the two board members hadn't raised the issue in any forum as far as Shawn knew. Shawn's boss kept saying he had the entire problem handled, but Shawn continued to explain his concerns. He made every effort possible to outline a reasonable decision process for site selection, scope, budget development, and project team development. Shawn identified other stakeholders impacted by the proposal as well as potential financial and contractual issues.

Shawn was especially concerned about the design of the structure that would surround and support the art installation. It had to be designed to work with the campus context, but even more importantly, it needed to be designed so that it was not a maintenance burden for the campus. Shawn had spoken to the other campus architects at another college and a university that had installations by this artist only to discover the artist had her own dedicated architect. This architect had only one mission and that was to make the artist happy. Evidently, the architect had a reputation as being difficult and not collaborative, an especially upsetting prospect for Shawn, who coached all project teams to be highly collaborative and to define shared values.

Broadsided

Six months later, Shawn's boss announced a contract had been signed with the artist. Stunned for the second time on this subject, Shawn asked why he had not seen a draft of the agreement. His boss lamely explained that the entire process had to be confidential until negotiations with both the artist and the donors were complete. As Shawn scanned the contract, he noted several clauses of concern. The site was confirmed in the document even though it had not gone through the board approval process. The perimeter measurements were part of the agreement, so Shawn noted the size of the installation had substantially grown. The agreement also required the use of the artist's architect for the project and required the College never to move or destroy the installation. Most alarming were the contractual terms for maintenance. The college committed to maintaining the installation and related structure to the artist's standards and closing the installation if the artist or artist's representative should ever deem it to be in substandard condition. These maintenance terms applied to all aspects of the installation including the technology and surrounding landscape. Although costs for the artist and architect had been included in the agreement, it seemed no one had developed a total project budget or estimated operational and maintenance costs.

Shawn had to laugh over the absurdity of the document and the decision process that had produced it. A number of the commitments made in the contract required board-level approval, but no such briefing, discussion, or approval had happened as far as Shawn knew. If it had happened, it occurred without critical financial and contextual information that would have shaped a very different contract. Apparently, Maryanne and the two board members had carefully manipulated the entire process. It was beginning to occur to Shawn that this project was not going to be conducted in a manner consistent with the principles and values he had understood to be important to the College.

The next months were a deeply frustrating for Shawn. A truthteller by nature, he found that the truth was the last thing that anyone involved with this project wanted to hear. No one believed the total project budget his team developed from the architect's project definition. No one wanted to hear about the impact to the campus master plan. No one wanted to consider how to pay for the total projected costs for operations and maintenance.

When the artist or donor wanted to add scope to the project, such as a complicated flashing detail or air conditioning the installation core, they didn't want to grapple with the fact there were no additional funds to pay for whatever they wanted. This lack of rational discussion was perhaps the hardest aspect of the two years it took to design and build the installation. Shawn was proud of building collaborative project teams including all stakeholders. These teams, shaped with common values, worked collaboratively to solve problems and meet quality, budget, and schedule goals—on every project but this one.

On this project, the artist and the architect refused to be part of any team-building process. The donors, who were also board members, also declined to participate. The usual decision discipline that was the hallmark of every other project at the College was utterly missing during this process. The architect and the donors challenged the pricing at every meeting. "There's no way it costs that much," was a phrase heard hundreds of times on the project. Project meetings were adversarial and unproductive. All Shawn could do was warn his boss, the vice president, the project would overrun the budget unless someone told the two board members to stop.

The frustration flowed onto the job site, where it became clear the artist's architect was unwilling or incapable of solving problems. The artist's specifications for visual performance were exact, but that did not necessarily translate to drainage, security, safety, and code requirements. Shawn had to hire another architect of record to address ordinary constructability problems and well as to ensure quality standards were met. All this added to the project cost. Of course, when the artist came on site, she would examine view corridors and insist that campus power distribution wires, transformers, traffics signs, and fire hydrants be relocated, to the dismay of the project manager and most everyone associated with the project.

Taking It on the Chin

Shawn did the best he could with the situation, although he was disillusioned and confused. He was sure that the campus could have worked with this artist in a very different way if the board had been clear with her on the values, principles, and priorities that drove their small, tuition-dependent college. However, it seemed when Maryanne and the two board members took over the process, these values, principles, and priorities went out the window. No wonder the artist and architect thought nothing of the College's needs. No one in a position to make a decision had ever communicated these needs as important.

As the installation drew near to completion and opening day was scheduled, Shawn sat through a few of the tests of the technology performance. He found the experience intriguing, but, he realized, he was jaded. This extravagant installation had come at too high a cost. The personal desires of artist and donors to make a mark on the campus had resulted in a multimillion-dollar loss for the campus and an eternal maintenance obligation for an institution that should be most invested in teaching and research. Of course, this installation would be a fabulous attraction for many visitors and there could be a positive cost-benefit analysis, if you considered the impact on the College brand.

However, Shawn suspected the damage to the College approval processes would be hard to repair. Deans had already lined up to find ways to befriend the right board member and were taking Maryanne's advice about how to bypass the system. The project delivery process had taken a hit as well, with trust dissolving on all sides. Significantly, the board, for the first time, was thinking about their own personal short-term desires rather than the long-term needs of the institution. Shawn's successors will not thank him for leaving the challenge of maintaining this piece of art no matter how extraordinary. He didn't like to think about all the resources that would be diverted to art installation maintenance rather than classroom renewal or technology upgrades. Shawn wanted to believe it was worth it, but in his heart, he knew his truth and his values.

Shawn contemplated the invitation once more. Then, with a slight smile on his face, Shawn tore the expensive paper into quarters, depositing the detritus into the circular file at the end of his desk. He grabbed his coat from the back of his office door, walked out to his car, and left the campus and the installation behind.

Embracing Risk and Contracts

6

Risk

I cannot even imagine what it would be like to be a goalkeeper at the world-class level in soccer. So many games are won or lost on a single goal, the stress must be immense. Penalty kicks are the worst, especially if they are coming after two hours of regulation and overtime with no score, as they did in the 1999 Women's World Cup. On a penalty kick, there is nothing between the goal and the player kicking the ball except the goalkeeper and this particular World Cup championship game would be decided by five penalty kicks by each team. As Jere Longman recounts in *The Girls of Summer*, goalkeepers "face worse odds than a carnival game, lucky to stop one attempt out of five." As these ultimate defenders stand before the nine-foot, ten-inch wide by six-foot, seven-inch high goal, the risk of missing all five saves is enormous.

Longman assessed the situation, "Hope for the keeper offered only three slim options—to guess beforehand and make a desperate lunge left or right, to maintain itchy patience and react after the ball was struck, or to read the shooter's body language on the approach and hope that the hip or knee or foot would betray some accidental intent."

It is the "inadvertent signal" or in poker parlance, a tell, from the other player that is the most valuable. If the goalkeeper can get any indication at all where the ball is going before the kick is made, she can act on that knowledge and better control the outcome.[1]

Two of the most challenging barriers to creating a Culture of Predictable Outcomes for the design and construction industries are misunderstanding risk and wrestling with poor contracts. Both conditions have flourished due to the lack of business education of many players in our industries. However, there are some basic principles related to each that can be easily revealed and understood. Knocking down these obstacles will go a long way to achieving the kinds of collaborative environments we need.

Risk Definitions

RISK, in the design and construction industries, is what you do not know and what you cannot control.

Risk is one of the most problematic and most misunderstood concepts in the design and construction industries, primarily because we have allowed others to define it for us. Dictionary.com defines Risk as "exposure to the chance of injury or loss; a hazard or dangerous chance," or, for the insurance world, "the hazard or chance of loss, the degree of probability of such loss, or the amount that the insurance company may lose."[2]

All this talk about injury or loss is a bit confusing in our world of design and construction because we are in the business of creating and building. Yes, we have the risk of injury on construction sites and of injury or loss from defective design or construction, but most of the risk we face every day on projects has little to do with this kind of danger. One interesting point to take away from this insurance definition, however, is that risk can be measured. Insurance companies have been quantifying risk for years.

The Insurance and Risk Management Professionals website defines risk as "Uncertainty arising from the possible occurrence of given events."[3] This definition is more helpful as it hits at the heart of our experiences in design and construction. Uncertainty and, therefore, unpredictability are what we face in our business. The more uncertainty we can remove from our projects, the more risk we remove. In other words, in our industries, **risk must be defined as what we do not know and what we cannot control**. The better the quality of knowledge and the better the team structures for executing projects, the lower the risk.

Risk Types

This discussion brings us to another important area of clarification regarding risk in the design and construction industries. There is a vast difference in value and definition between risk avoidance, risk management, and risk reduction as follows:

Risk Avoidance is what lawyers, insurance companies, and even the AIA have been encouraging the designing construction industries to do for years. Avoiding risk means avoiding responsibility and avoiding opportunity. Risk avoidance handcuffs team members and creates animosity. Risk avoidance crushes innovation. You may have heard the aphorism "there is no reward without risk." This saying is mostly true, although I think it would be wiser to say there is no reward without responsibility. With responsibility comes risk. Once you have that risk, you have two choices: you can manage it or reduce it.

Risk Management is the process of carefully identifying risk and putting plans in place to manage the consequences of that risk (not the same as avoiding the risk). For example, you may take on a new team with some members you do not trust. To manage the risk, you might put a trusted employee on the team and increase the frequency of check-in meetings. Typical risk management strategies include insurance and bonding, although these structures are not nearly as useful as they should be because they are designed by people outside our industries that want to take advantage of our naivete.

Risk Reduction is the process of accepting risk, then methodically reducing or eliminating risk by increasing *knowledge or control* over that risk. Collaborative environments reduce risk significantly more than non-collaborative environments by providing greater transparency and greater knowledge sharing, thus improving problem-solving and accelerating reaction to the market (see Figures 6.1 and 6.2). For example, Sutter Health adopted Target Value Design in 2004, a highly collaborative process of managing a project budget transparently and ideating cost reduction continuously.[4] Risk reduction also plays a part in decision risk. The better the quality of the "decision-ready" information and the more timely that decision is made, the more knowledge and control you have over the decision and the lower the risk of that decision. The most important takeaway here is to remember when you actively reduce risk on any project, you build value on that project—value for every team member.

The **Theory of Expanding Risk** is a concept I talk about in *The Owner's Dilemma*. Just as you can create environments, processes, and contracts that reduce risk, project stakeholders can significantly, even exponentially, increase risk on a project.[5] Beware of processes that create bureaucracies, opaque decision-making, or undermine team values. The design and construction industries must move quickly away from risk avoidance, so we may enter an era of extreme collaboration, partnership, and innovation. There will be some risks that must be managed, but many risks can be reduced or eliminated through collaboration and innovation if we do not let attorneys and insurers get in our way.

Contracts

Don't get me wrong, just like any profession, there are good attorneys and there are bad attorneys. Perhaps I should say, there are attorneys that deeply

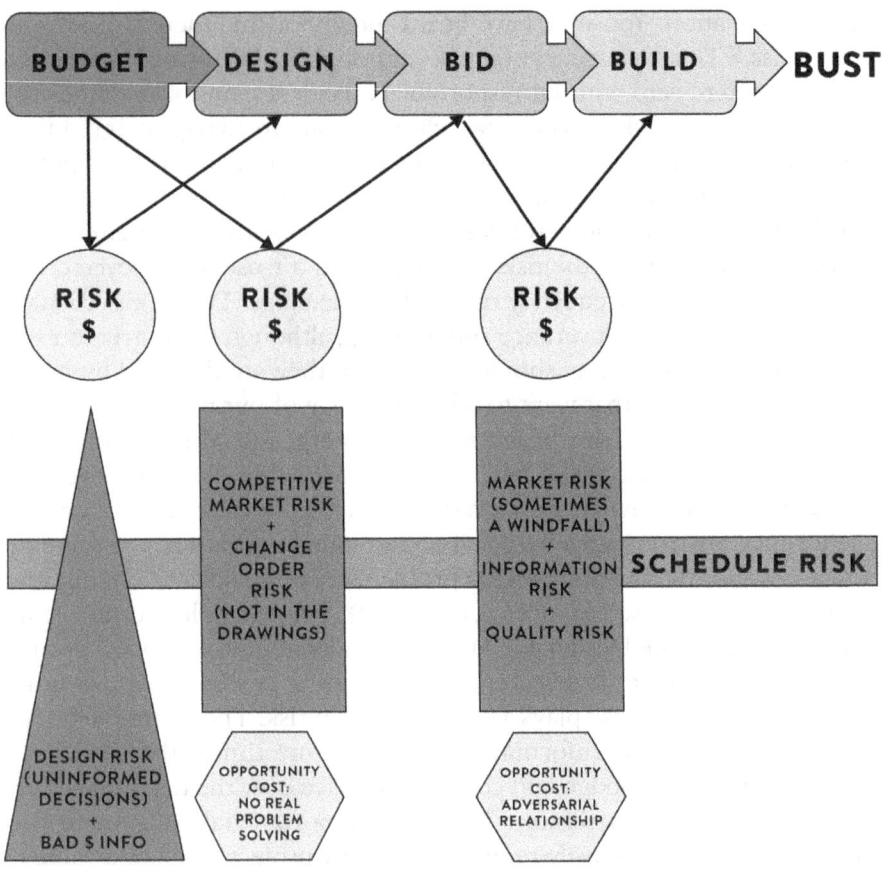

Figure 6.1 Design/Bid/Build Risk Assessment – The traditional processes embeds unmanageable risk throughout the process increasing the probability of "Busting" the budget and schedule; as shown in *The Owner's Dilemma*

understand what we do and attorneys that have absolutely no idea at all. It's the attorneys in this second category, often well meaning, that have caused so much damage to our industry, pushing us into silos and prioritizing risk avoidance above all else. Bill Shankly, Scottish footballer and manager, once said something that hits home to me, "The trouble with referees is that they know the rules, but they don't know the game."[6] Understanding the letter of the law or insurance—I'll include insurance brokers here—doesn't mean an attorney or broker knows how to leverage that knowledge to benefit the business of design and construction.

I cannot place all the blame on the lawyers and insurance brokers. Frankly, there is a long-standing problematic relationship between the design and construction industries and its contracts. Just as with risk, there is a general

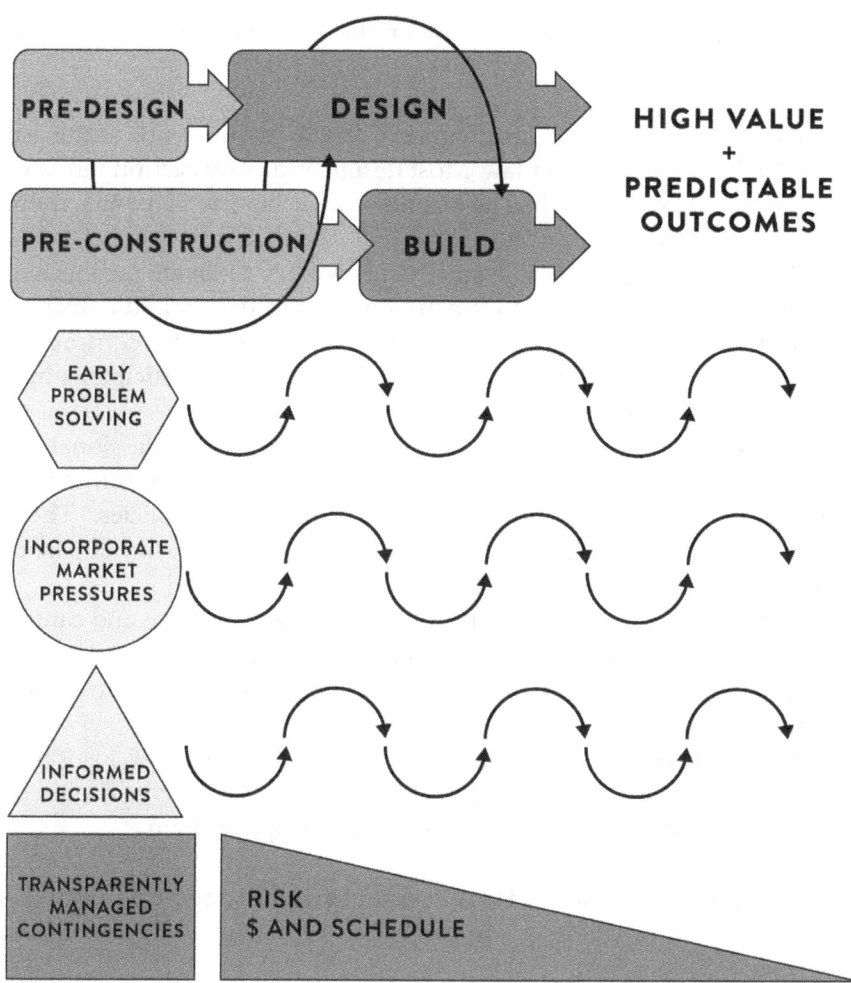

Figure 6.2 Collaborative Process Risk Assessment – A process where risk can be identified, quantified and managed, can help lead to predictable and high value outcomes; as shown in *The Owner's Dilemma*

misunderstanding of what role contracts can and should play within a project. This fundamental misunderstanding has created a set of contractual relationships in our industries that make us act in ways we do not want to and, often, squelch innovation. When contracts lead our professional relationships rather than a culture of predictable outcomes, we get stuck in destructive, value-sucking patterns. These types of contracts reduce value for every member of the team.

I must warn you now. I am NOT a lawyer. I am an experienced practitioner. Always consult a trusted attorney before you sign any contract.

However, I suggest professionals consider the following as they develop contractual relationships in the future:

1 **Context**—Understand the legal context in which you work and learn about the basics of contract law. Most design and construction industry professionals have never had a business law class, which puts these professionals at a disadvantage.
2 **Siloed Relationships**—Most legal advisers will try to divide responsibilities and liabilities between contractual parties in design and construction contracts with the intention to reduce risk. Instead, this contractual device most often results in increased risk by pushing participants into opaque adversarial relationships.
3 **Unintended consequences**—Many legal and insurance professionals do not understand our industries or the work we do, causing unintended consequences while working on our contracts and policies. These include remedies such as liquidated damages, adding risk for all parties by increasing adversarial oversight.
4 **Acting badly**—Contracts do not force people to misbehave and cannot compel others to act well. People choose their actions based on their individual level of understanding of the consequences of their actions and their personal values.
5 **Reversal of Roles**—You do not work for legal advisers or insurance advisers. They work for you. Legal and insurance instruments should be created to support your business goals, not to impede them.

Appreciating these dynamics and turning them around to serve a Culture of Predictable Outcomes will help you meet your project goals from the start of the project.

1 Context

Understand the legal context in which you work and learn about the basics of contract law so you can flourish in these complex industries. Most design and construction industry professionals have never had a business law class, which puts these professionals at a disadvantage.

Definitions

I have always said that definitions are everything, so let's start clarifying this problem with the definition of the word "contract" from a business law text.

A contract is "a promise or a set of promises for the breach of which the law gives a remedy, or the performance of which the law in some way recognizes as a duty." Put simply, a contract is an agreement that can be enforced in court. It is formed by two or more parties who agree to perform or to refrain from performing some act now or in the future.[7]

I will phrase this even more clearly. Contracts are tools meant to serve the parties participating in the contract. Somewhere along the way, that concept has been somewhat forgotten in the design and construction industries. Contracts and those writing contracts have taken charge. Required contract language took control of our lives, our professions, and the way we do our work.

It's important to remember the role contracts play in our industries. We do not have a series of simple transactional relationships. Each project is comprised of dozens, sometimes hundreds of contracts. Industry attorney Christopher Noble observed,

> In the future, contract and licensing agreements will be more important than ever. Written contracts play a greater role in design and construction than in almost any other industry, in part because our industry consists of many independent participants who come together on a project-by-project basis rather than of large integrated entities with stable, continuing relationships.[8]

Contracts are supposed to be an agreement meant to benefit both parties—to allow you to access something you want or need—something you would not have access to without the contract. Yet, contemporary design and construction contracts, regardless of, or perhaps because of, the protections written by lawyers, now have the unintended consequences of forbidding us from collaborating the way we wish, taking responsibility the way we wish, or even innovating the way we wish. Happily, this dynamic is changing on some projects, but far too slowly.

A Little History

I was fascinated by a story told by Chuck Thomsen and Sid Sanders in their book, *Program Management 2.0*, as they discussed the history behind the architectural contract and scope of responsibility.

> When the craftsman/builder/architect moved out of the mud and rain, received degrees from universities, established associations like

the AIA, obtained licensing and became a professional, the concept of a Standard of Care replaced the concept of a defect-free building. AEs argued that to err is human and their responsibility should not be perfection. Instead, their work should only meet the standard of their peers. Although circumstances, jurisdictions and the predilection of judges differed, most courts agreed that an architect did not guarantee perfect plans, a perfect building or perfect supervision that would deliver a defect-free project.[9]

The narrative describes a slow retreat in legal responsibility from Standard of Care to reasonable care and reasonable powers of observation as an architect oversees construction.

The authors discuss the responsibility of the architect as professionals to exercise judgment, accepting the limits of what can be known on a construction site.

It's the recognition under common liability law that professionals (doctors, lawyers and architects) are exercising learned judgment. That judgment is based on experience and a body of knowledge recognized by degrees and licenses. They deal with conditions not totally knowable or under their exclusive control.[10]

Builders felt the weight of liability being transferred to them as the experts on the job site and sought protection. "Their position was that they weren't liable if they built what was designed and specified."[11]

And so, the struggle began. There was no way to resolve this tension between two separate contracts with the owner, especially with the architect avoiding responsibility for a defect-free building, and a builder accepting responsibility only for plans and specifications provided.

Owners Have All the Risk

I stated in *The Owner's Dilemma* in 2010 that the owner accepts most, if not all, of the risk of a construction project.

Owners are a powerful lot, not just because they initiate and pay for projects, but also because they have always taken the lion's share of the risk. I include developers who are both serial builders and owners in this category. Owners choose project teams and set the tone. Owners determine sustainability and performance goals. Owners select the delivery processes. Owners make decisions.[12]

I also stated the owner was happy to spread that risk around to the project team under certain circumstances. Many disagree with my contentions about owners' risk. However, there is legal grounding for this argument considering the architect is not required to deliver a "defect-free" building, and the contractor is only required to deliver projects based on the plans and specifications provided by the architect. Thomsen and Sanders address this,

> Early in the 20th century, the United States Supreme Court held that since a builder agrees to build according to plans and specifications furnished by the owner (and it can't be shown that the contractor knew that the plans and specifications would produce a defect), the contractor is not responsible for the consequences of defects in the plans and specifications. If a builder builds a flawed design as defined in the plans and if the architect could demonstrate that he or she had used the "care employed by reasonably prudent professionals practicing in the same field in the same area," the owner is left with the cost of correction. Even though there is a mistake, the AE is not liable, the builder is not liable, and the owner must pay. Owners can never shed all risk. Initially, all risk on a building project lies with the owner.[13]

Recognizing and being sensitive to the owner's risk position with regard to legal precedent is essential as you work your way through contracts.

What You Don't Know

Historically, our industries' contractual and risk management instruments have been primarily shaped by those that do not know the industry as well as we do. We have relied on legal professionals and others to tell us what to do without understanding the impact they can have on our professional lives and our projects. There is a good deal to worry about in today's professional and legal environments, as noted by Julia Mcloughlin, RIBA councillor,

> During recent years the profession has become much more competitive and high risk, with far greater potential for things to go wrong with clients, fees and building contracts. So, in this highly litigious society how can we ensure we avoid the many potential risks and pitfalls in our architectural careers?[14]

I suggest instead that design and construction professionals concentrate on opportunities rather than what can go wrong. James Burgoyne, lawyer and insurance broker, advocates, "It is essential that the modern practitioner has a working legal, insurance, risk management and information management knowledge. These are key supports to their business—the economic conduit for their passion and expertise."[15] Learn to ask questions to understand how these instruments can better serve your business. Question the foundational concepts if they do not serve you. If you want to deliver a defect-free modular home, design a contractual instrument that will serve that product, and then negotiate the associated value. Learn what you do not know. You do not have to become an attorney to become knowledgeable about the concepts. Read your contracts and other instruments carefully and ask questions. I always say, "Ask why five times"—until you understand why that instrument is written as it is and what it is intended to do.[16]

2 Siloed Relationships

Most legal advisers will try to divide responsibilities and liabilities between contractual parties in design and construction contracts with the intention to reduce risk. Instead, this contractual device most often results in increased risk by pushing participants into opaque adversarial relationships.

In addition, this contractual separation of responsibilities has the unintended consequence of preventing the team from delivering the best project possible for the best value.

Separation of responsibilities and liabilities seemed a perfectly rational approach to contracting in the early twentieth century, as noted by IPD specialist and lawyer Howard Ashcraft,

> traditional construction contracts confine parties to their assigned roles and similarly segregate liability. This was a feasible risk management approach because the economic-loss doctrine applicable in many states prevents an injured party from recovering pecuniary losses unless the party has a contractual relationship to the defendant.[17]

However, like the initial onset of competitive bidding, although the initial intentions were great, the unintentional consequences of separating contracts and responsibilities and limiting liability on a design and construction project became problematic. These contracts began to produce unexpected complications, especially as building projects became more complex. As Rodd Merchant notes,

> Today nearly all methods of building project delivery are structured, legally and otherwise, to sever the interrelated functions of design, material procurement, handling, fabrication, shipping, installation, and operation. Current model contracts are purposefully structured to limit the risk and exposure of each player in the process, which in effect, detracts from the completed work at the owner's expense. As a result, participants tend to work in silos focused solely on their assigned scope, ignoring the interrelated activities of others.[18]

Not only do these contracts constrain team members, limiting their scope of work, but they contribute to frustrations and manufacture problems. Peggy Deamer observed this dynamic in *Building (in) the Future*, "Contracts memorializing business practices designed to limit liability, already having shown themselves to increase antagonism and constrain design's social reach, had been clearly proved to be impediments to disciplinary empowerment."[19]

If I have made a sufficient argument in support of high-performing collaborative teams in this book, then it is not a great leap to expect that contracts should support those collaborations. Contracts should not create risk in sharing knowledge or relying on information from other team members. Contracts should not artificially create gaps in project scope by forcing each discipline into a narrow silo even when complex problems arise. These practices increase risk by preventing decisions from being made as quickly as possible and in the most informed fashion by preventing inclusive problem-solving.

3 Unintended Consequences

Many legal and insurance professionals do not understand our industries or the work we do, causing unintended consequences while working on our contracts and policies. These include remedies such as liquidated damages adding risk for all parties by increasing adversarial oversight.

The architectural community was never so poorly served as when it was ordered to stand down from asbestos removal consulting by the insurance industry, a remarkably low-risk endeavor. Architecture firms could have made lots of bread-and-butter profits by assisting owners during the late decades of the twentieth century. However, insurance companies and lawyers ordered firms to stay away, and architects did stay away. This example demonstrates how many legal and insurance experts do not understand the design and construction industries and therefore provide poor advice.

Another example is the hiring of a Geotechnical Engineering consultant (Geotech). Traditionally, Geotechs work directly for the owner, and the structural engineer works for the architect who works for the owner. Both are needed to design and inspect the foundation. Both are liable for foundation design and performance. However, they do not work together for the architect because some insurer and some lawyer decided that evaluating the condition of the soils was too great a liability for the architect. This analysis makes little or no sense, especially in a collaborative environment with greater value created when the two consultants work together. So, why does this legal and insurance barrier still exist?

"Shared savings" on contractor-owner contracts is another popular concept embraced by lawyers and some owners but with significant unintended consequences such as incentivizing the contractor to overprice initially and setting up an adversarial relationship between the contractor and architect, who is left with the responsibility for checking quality, throughout the project with no incentive.

These examples are just a few contractual or risk avoidance measures that have resulted in unintended consequences largely because those persons recommending the measures do not understand the work we do. Traditional project management, legal, and insurer guidelines put as many protections in place as possible. According to some legal experts, checks and balances, separations, and clearly defined responsibilities should create a perfect project. As Thomsen and Sanders note, it's not that easy,

> In the past, management practices have viewed organizational authority, precise contracts, detailed schedules and legal recourse for non-performance as the appropriate tools for knitting together such ad hoc organizations. The theory is that if each does its work satisfactorily, as specified and scheduled, the result will be OK. It's a logical theory but disappointing in practice. The problem is that there are always changing and unexpected circumstances. And we live in an imperfect world where everyone makes mistakes. That breeds conflict and legal action. Legal action ricochets. If one party sues another, the defense is to find the plaintiff's mistakes and counter sue. The conflict spreads from there. Since everyone has made mistakes, everyone is open to blame.[20]

Our project delivery environment is complicated and getting more complicated all the time. Complex, changing business environments need resilient teams. Resilient teams need contracts that permit them to be nimble and flexible. Checklists won't work. Reasonable autonomy guided by values and principles will work. It's a new day regarding contracts, one that must be led by knowledgeable industry professionals and legal professionals interested in helping us work together.

When legal professionals do not understand what we do, they create legal issues related to knowledge and digital sharing first with CADD and, later, BIM, building additional barriers to collaboration and creating redundant work. Howard Ashcraft tells the story,

> as parties began to exchange computer-aided design and drafting (CADD) data electronically, questions arose concerning the ability of the receiving party to rely on the information and the liability of the transmitting party for its accuracy. Under Restatement of the Law Second: Torts §522, a person who negligently supplies incorrect or misleading information to a person who justifiably relies on the information is liable for any financial loss sustained. Free exchange of CADD data thus raised concerns of expanded liability to third parties. In response, parties reinforced the contractual walls by disavowing responsibility for the transmitted information. But this approach leads to inefficiency, misunderstanding, and poor outcomes, which inevitably leads to claims. And project efficiency requires reliable communication. A change is needed.[21]

Happily, Ashcraft notes the AIA issued the Digital Data Licensing Agreement and the Digital Data Protocol Exhibit in 2007, explicitly permitting "reliance on digital information for permitted project purposes." In addition, the Associated General Contractors (AGC) adopted the Electronic Communications Protocol Addendum. According to Ashcraft,

> There are significant differences between the documents: the Electronics Communications Protocol Addendum uses a checklist and mechanics approach to data transfer, whereas the Digital Data Protocol Exhibit focuses on the procedures, formats, and purposes of the data exchanges through the Project Protocol Table. But their differences are dwarfed by their similarities. Both assume that information must be freely transferred and that the receiving party may rely on the information in executing its project responsibilities.[22]

There has been progress, but many more challenges lay ahead.

The ever-increasing complexities of the industry are daunting, and the legal instruments aren't up to the challenges. Thomsen and Sanders observe,

> The contract structure for most projects is vertical: consultant to architect to owner—and manufacturer to supplier to subcontractor to CM to owner. Yet many of the commitments are horizontal or diagonal across contract lines. For instance, the structural engineer must provide the mechanical engineer with the framing plan so the mechanical engineer can lay out ductwork, yet there may be no contract between the two. The members of the project team must give and extract commitments to and from others that aren't explicitly contractual. Most commitments cascade through the project.[23]

Many industry lawyers still worry about letting the client rely on architects' professional work. They lose sleep over the appropriate distribution of responsibility and are more worried about the potential bad actors than worry about the lost opportunities of keeping our industries handcuffed. Thankfully, a few extraordinary attorneys, such as Howard Ashcraft and Will Lichtig, are more interested in setting us up reactively for a fight.

4 Acting Badly

Contracts do not force people to misbehave and cannot compel others to act well. People choose their own actions based on their individual level of understanding of the consequences of their actions and their personal values.

A couple of years ago, I was speaking out on the West Coast. I finished my talk in a large ballroom. The lights were set so that I could see the faces in the audience despite the bright light shown on the stage. I was rewarded by a high level of energy and engagement in those faces and was, therefore, very glad we had time for questions. The first question, though troubling, was not unusual,

> We are engaged in a series of design-build projects where the contractor is the lead. We are getting killed by the consequences of a contract that allows the contractor to change what he likes, and then we have to adjust our design. The contract makes him do this: What can I do?.

I hear some version this question time and time again. Versions include, "The contract made me do it!" or "The contract made him or her do it" or the real classic, "The contract won't let me do it."

We all know that during the day-to-day struggles of a project, team members' actions have little to do with what the contract says. Practicing transparency of information would not be inhibited by a contract unless a nondisclosure agreement is included. The contract rarely drives refusal to collaborate in a problem-solving effort. Making changes without collaborating with the team to understand consequences is also unlikely to be **required** by the contract. I have never seen such a clause.

Frankly, contracts drive few of the behaviors you see in projects. Individuals choose their own actions based on their individual level of understanding of the consequences of their actions and based on their personal values. Admittedly, team members often misunderstand contract requirements due to the way they are written or because they have absorbed them through hearsay versus reading the contract. To truly control all behavior, contracts would have to be millions of pages long. Thomsen and Sanders note,

> As long as multiple organizations have interests in a team and those organizations are doing at least some of the work independently, unaligned self-interest will exist. There will never be a contractual vehicle that will replace the need for professionals who have their hearts in the right place.[24]

The corollary is, although contracts cannot and do not solve every behavioral problem on a project, nor do they create every problem. Contracts can inhibit or support behavior. Design-bid-build contracts incentivize contractors to bid low, hide information, take advantage of mistakes in the architect's drawings, and rely on change orders to make a profit. Not all contractors do, but some do, and the contract allows it to happen.

On the other hand, project teams seeking a value-generating collaborative process might embrace an IPD contract that would incentivize collaborative behavior. That team might have practiced collaboratively anyway. However, the contract takes away barriers that might inhibit optimum collaborative performance. The authors of *Integrating Project Delivery* found this to be true in their research, "until Sutter Health solidified their approach to contractually bind the parties to a common goal with shared risk and reward, they did not get the behavior they sought."[25]

Howard Ashcraft calls for a more aggressive buy-in by team members to make the IPD contract work. He notes,

> IPD requires even more profound changes in belief and behavior. IPD is a trust-based project delivery methodology and will not succeed,

regardless of structural changes, unless the team understands why IPD works.[26]

Therefore, although contracts rarely make people behave badly or make them behave well, contracts can support and encourage desired behavioral outcomes. A carefully crafted contract is an important supportive tool for a Culture of Predictable Outcomes.

5 Reversal of Roles

You do not work for legal advisers or insurance advisers. They work for you. Legal and insurance instruments should be created to support your business goals, not to impede them.

Many attorneys and insurers act as if Owners, Architects, Engineers, and Contractors work for them and that we should all follow their direction without question. Let me be clear: design and construction industry professionals do not work for attorneys and insurers. Industry professionals and owners pay attorneys and insurers for their services. Attorneys and insurers provide services to industry professionals and owners. Attorneys and insurers should be developing contracts and insurance instruments that serve the needs of industry players. Industry professionals need to

Do architects know any short words?

understand this power relationship so we can gain greater freedom to collaborate, innovate, and deliver better projects.

There are exceptions. For example, in my role directing university construction, I am used to working with a general counsel who has a broader responsibility to the institution. In this case, developing a long-term partnership to inform the in-house legal adviser about the opportunities for collaboration and innovation is a good strategy. Another exception is an attorney working with a team in partnership, proactively building a contract to support collaboration such as in IPD. In this case, the partnership is with the entire team, including the legal adviser. In all cases, it is imperative that the tail not wag the dog. We need legal advisers and insurers to find a way to help us support a Culture of Predictable Outcomes.

Solutions

In 2017, McKinsey did a series of articles on the poor performance of the construction industry. One of the reasons identified for poor efficiency was the contractual frameworks currently used within the industry.

> Many in the industry shared case studies demonstrating that when interests are aligned and aimed at well-defined outcomes, projects are more likely to meet schedule and cost targets. To align interests, the industry must move away from the hostile contracting environment that characterizes many construction projects to a system focused on collaboration and problem solving.[27]

To accomplish the goal of moving away from traditional contracting, owners and project teams must consider a few concrete recommendations:

1 **Start the conversation early and collaboratively about expectations and barriers**. For example, if standard contracts must be used because of regulatory requirements, discuss the unintended consequences of applying contract solutions rather than collaborative solutions when a project faces challenges. If possible, provide language permitting alternative solutions. Also, discuss institutional barriers to success that may impact pricing or behavior. Are there cultural differences between organizations? Are there hidden financial issues that will come back and create discord later on? Is the facilities department notoriously hard to work with? How can that be mitigated?

2 **Negotiate key elements of the contract as a team**. In the report *Integrated Project Delivery: Action Guide for Leaders,* a chapter titled "Path to Contract" outlines a process to develop a contract reflecting project and team goals. "The contract workshop is intertwined with the team alignment process. Separating them creates the distinct likelihood that the contract will not be consistent with the IPD team's values and goals and the team will not understand the structure they must work in. Moreover, contract negotiation is often simpler and less contentious when those responsible for contract negotiation—legal, procurement, and risk management personnel—have a solid understanding of IPD principles and can see for themselves how a team of disparate companies can work collaboratively on developing common goals and values."[28] This alignment strategy is an excellent approach, even if your team is working on another type of collaborative delivery process. Aligning the financial model and agreement also makes sense.

3 **Create incentives in a manner that is fair and directly aligned with goals**. I have always believed in creating teams where it is possible and probable that every member is successful at the end of the project. At Rice, I was not able to create a comprehensive incentive structure, but it is something I wished I had accomplished. Thomsen and Sanders advocate, "Financial rewards are a useful tool to help align a project team to project goals. To set such awards, the owner must stipulate them and define the process and metrics for measuring them. The goals should be the same for each of the core team members and, in general, they should share the rewards proportionate to their participation in the project."[29] Although IPD seems to be the most proven at this writing, there are different structures for this process.

4 **Be innovative from the start and leave room for more innovation**. Technology, delivery processes, and construction methods are changing rapidly at the start of the third decade of the twenty-first century. Our contracts must keep up. Standard contracts must be flexible and updated quickly and easily, unlike the typical AIA models. New contracts must anticipate procurement changes, as well as the fluidity of design and design responsibility. Allowing this flexibility in a contract is hard, but it has to be addressed. Rodd W. Merchant outlines a simple example in *Building (in) the Future,* "Under this design-build/strategic-alliance model, the fabricator still holds the prime contract for the steel frame, but the engineering is provided under the subcontract to the fabricator. For this model to work, the steel subcontract must be awarded early on, allowing participation and input by all team members. Since the engineering is provided by the fabricator, the preferences of the detailer, mill

provider, erector, and other product suppliers can be incorporated into the design. The team internally controls activities such as coordination, schedule milestones, and submittal and shop drawing processes. During construction, the team also handles any problems arising from fit-up or fabrication errors, eliminating the need for requests for information or nonconformance requests on the part of the general contractor and/or architect."[30]

5 **Follow-through on behavioral commitments made during workshops and built into the contract**. Falling back into traditional contract mode will push you right back into old-school design and construction contract behavior regardless of what the contract says. Practice all the elements of the Culture of Predictable Outcomes for best results.

Summary

Understanding risk and contracts is a foundational requirement for setting a Culture of Predictable Outcomes in place. Learning that risk can be reduced through collaborative processes changes mindsets and makes innovation possible. Do not forget, **risk in the design and construction industry is what you do not know and what you cannot control**. Both knowledge and control are supercharged in collaborative environments.

Change your power relationship with contracts and make sure they serve your business and your teams as a tool. The right contracts can support the goals you have for your collaborative processes. Dream big and push back if you are not getting the contracts you need to proactively solve problems and create great environments. Seek legal and insurance advisers that want to know and understand the business you want to practice. Warren Buffet once said, "It is impossible to unsign a contract, so do all your thinking before you sign."[31] I say you should also understand what you are thinking about. Do not be afraid of risk and contracts. Make them your friends and they will help you what you want to do.

Notes

1 Jere Longman, *The Girls of Summer: The U.S. Women's Soccer Team and How It Changed the World* (Pymble, Australia: HarperCollins e-books, 2000), 85–86.

2 "Risk," Dictionary.com, June 15, 2020, www.dictionary.com/browse/risk?s=t.

3 "IRMI Glossary: Risk," Insurance and Risk Management Professional website, June 15, 2020, www.irmi.com/term/insurance-definitions/risk.

4 Glenn Ballard and Peter Morris, "Maximizing Owner Value through Target Value Design," *AACE International Transactions*, January 1, 2010, 356–357.

5 Barbara White Bryson and Canan Yetmen, *The Owner's Dilemma: Driving Success and Innovation in the Design and Construction Industry* (Atlanta: Ostberg Library of Design Management, Greenway Communications, 2010), 22–24.

6 M. Prefontaine, *The Big Book of Quotes: Funny, Inspirational and Motivational Quotes on Life, Love and Much Else* (n.p.: MP Publishing, Kindle ed., 2015), 297.

7 Kenneth W. Clarkson, *Business Law: Text and Cases* (Australia: Cengage Learning, Kindle ed.), 225.

8 Peggy Deamer and Phillip G. Bernstein, eds., *Building (in) the Future: Recasting Labor in Architecture*, (New York: Princeton Architectural Press, 2010), 175.

9 Chuck Thomsen and Sid Sanders, *Program Management 2.0: Concepts and Strategies for Managing Building Programs* (rev. ed.) (McLean, VA: CMAA, 2011), 39.

10 Ibid., 40.

11 Ibid., 42.

12 Barbara White Bryson and Canan Yetmen, *The Owner's Dilemma: Driving Success and Innovation in the Design and Construction Industry* (Atlanta: Ostberg Library of Design Management, Greenway Communications, 2010), 12.

13 Chuck Thomsen and Sid Sanders, *Program Management 2.0: Concepts and Strategies for Managing Building Programs* (rev. ed.) (McLean, VA: CMAA, 2011), 42.

14 Alan Jones and Rob Hyde, eds., *Defining Contemporary Professionalism: For Architects in Practice and Education* (London: RIBA, 2019), 172.

15 Ibid., 51.

16 Barbara White Bryson and Canan Yetmen, *The Owner's Dilemma: Driving Success and Innovation in the Design and Construction Industry* (Atlanta: Ostberg Library of Design Management, Greenway Communications, 2010), 149–150.

17 Peggy Deamer, Phillip G. Bernstein, eds., *Building (in) the Future: Recasting Labor in Architecture* (New York: Princeton Architectural Press, 2010), 149.

18 Ibid., 160–161.

19 Ibid., 20.

20 Chuck Thomsen and Sid Sanders, *Program Management 2.0: Concepts and Strategies for Managing Building Programs* (rev. ed.) (McLean, VA: CMAA, 2011), 390.

21 Peggy Deamer and Phillip G. Bernstein, eds., *Building (in) the Future: Recasting Labor in Architecture* (New York: Princeton Architectural Press, 2010), 149.

22 Ibid., 150.

23 Chuck Thomsen and Sid Sanders, *Program Management 2.0: Concepts and Strategies for Managing Building Programs* (rev. ed.) (McLean, VA: CMAA, 2011), 418.

24 Chuck Thomsen and Sid Sanders, *Program Management 2.0: Concepts and Strategies for Managing Building Programs* (rev.) (McLean, VA: CMAA, 2011), 390.

25 Martin Fischer, Howard Ashcraft, Dean Reed, and Atul Khanzode, *Integrating Project Delivery* (Hoboken, NJ: Wiley, 2017), 23.

26 Peggy Deamer and Phillip G. Bernstein, eds., *Building (in) the Future: Recasting Labor in Architecture* (New York: Princeton Architectural Press, 2010), 152–153.

27 Filipe Barbosa, Jonathan Woetzel, Jan Mischke, Maria Joao Ribeirinho, Mukund Sridhar, Matthew Parsons, Nick Bertram, and Stephanie Brown, "Reinventing Construction: A Route to Higher Productivity," McKinsey Global Institute, February 2017, www.mckinsey.com/~/media/McKinsey/Industries/Capital%20Projects%20and%20Infrastructure/Our%20Insights/Reinventing%20construction%20through%20a%20productivity%20revolution/MGI-Reinventing-Construction-Executive-summary.ashx, 3.

28 Charles Pankow Foundation, Center for Innovation in the Design and Construction Industry, Integrated Project Delivery Alliance, *Integrated Project Delivery: An Action Guide for Leaders*. Accessed print on demand May 26, 2020, 31.

29 Chuck Thomsen and Sid Sanders, *Program Management 2.0: Concepts and Strategies for Managing Building Programs* (rev. ed.) (McLean, VA: CMAA, 2011), 371.

30 Peggy Deamer and Phillip G. Bernstein, eds., *Building (in) the Future: Recasting Labor in Architecture* (New York: Princeton Architectural Press, 2010), 164.

31 Mary Buffett and David Clark, *The Tao of Warren Buffett* (New York: Scribner, 2006), 8.

Chapter Supplement—6
Risk, Rapport, and Reversing Roles, or Inviting the Fox into the Henhouse

A Good Idea

Phil Swain, AIA was happy to volunteer when the Catalina Foothills School District came to him several years ago asking for assistance modifying building permit statutes at the state legislature to reduce the cost and bureaucracy associated with school renovations. During the process of explaining the case for reducing building permit requirements to the state legislature, Phil and his team came to a remarkable realization. *Architects are ultimately responsible and liable for the quality, safety, and code compliance of their designs regardless of whether a permitting authority reviews the construction documents or not.* So, why not allow architects to take control of the approval and inspection processes under certain conditions?

Phil Swaim had no problem with accepting the responsibility of his profession along with the risk integral with that responsibility. He remembered the dark days of the 1980s and 1990s with the risk rollback initiated by the AIA, the lawyers, and the insurers worried about liability. Under the guise of protecting the profession, these entities encouraged architects to step back from being too involved in construction, especially from inspection of construction or determining "means and methods." As a result of this advice, the halcyon days of the architect as master builder were now long gone. Alternatively, Phil asserted,

> We consciously decided here at Swaim [Associates Architects] to go in the opposite direction. We would much rather be involved wholeheartedly with our construction partners on the day-to-day level, whether we're getting paid on the level we need to or not. It's much healthier. We learn more. It means less risk to us [because we're] involved.

Remarkably, the Arizona State Legislature passed a bill, surpassing the architect's and School District's expectations, allowing any county with a population over 500,000 and less than 1.5 million to build public schools without building permits. The bill gave the architects full responsibility for code compliance and inspections. Phil Swaim was delighted with this opportunity, "It's been good for our teams to be more responsible and more knowledgeable instead of relying on another entity to look after the permitting and inspections, and then complaining about how long they take."

The Challenge

Swaim Associates Architects had been doing their own code reviews and building inspections for all their Pima County school projects for nearly eight years when the Mica Mountain High School project came along. On a fifty-seven-acre campus, the comprehensive high school would be delivered to the Vail School District in quick phases. Challenged by both schedule and budget, the project would open to 450 or more students in the first phase, including a student union, two classroom buildings, a staff house, arts and auditorium building, two athletic buildings, the athletic facilities, and a pre-school. Later a dining room and kitchen, culinary arts, engineering, and career technologies facilities would be added. The entire campus would be supported by solar canopies providing shading and power. The first eight buildings had an eight-month construction schedule.

Fred Knapp of Core Construction, the contractor selected for the project, explained that Core was hired as a Construction Manager at Risk (CMAR) and was included in the project team early in the programming phase. The construction team, including major subcontractors, provided design-assist services. The team could tell from the very beginning that the budget was going to be an enormous challenge. Therefore, having subcontractors on board early provided the opportunity to work with the architects on building types, structures, and envelopes. The team was able to create a mix of materials and building methods that, eventually, looked like it might meet the budget and schedule challenges of this vast and complex project.

However, on this project, Phil Swaim was concerned that his firm would not have the capacity to oversee all the code and inspection responsibilities of all the buildings on the campus. Yet Phil did not want to give up ownership and responsibility for code review and inspections. He also had no desire to burden the project with the time and cost of a traditional building permit and inspection process from Pima County, although the County had made great strides in streamlining their systems in recent years. The team considered using an inspection consultant for the daily inspection processes, but that was an expensive option.

Role Reversal

As Fred Knapp remembered, it was Phil Swaim who came up with the solution, to reach out to Pima County and ask the county to work for the project team, as a consultant to deliver limited code review and construction inspections. Fred was unsure how the Pima County officials would take the idea, "Phil was the one who reached out to them. We just sat in a conference room, and talked through the prospect of Pima County doing our inspection as a consultant." Phil laid out the proposal asking Pima County to conduct two plan reviews of the construction documents and to conduct all construction inspections.

As the licensed architect representing Swaim and Associate Architects, Phil Swaim would remain the Authority Having Jurisdiction (AHJ) but would carefully consider all concerns raised by Pima County during the code reviews. Pima County would provide a single inspector for the entire duration of the Mica Mountain High School project, and Core Construction would provide an office for the Pima County inspector in the construction trailers. The inspector could use the office as a home base to serve all east side projects for Pima County as necessary. However, Mica Mountain High School would be the inspector's priority.

This proposal was admittedly unusual. As Fred Knapp remembered the proceedings,

> I think [the County Officials] could see the benefits of the additional revenue, knowing that once that school construction legislation was passed, they had to figure out a different way to serve the community. So they stepped up and provided an option.

For all proposed services, Phil Swaim and Pima County negotiated a fair price for both entities, also representing a reasonable price for the school system, the project owner. If all went as planned, the solution would work well for the budget and help the team meet the schedule.

However, Phil Swaim could not help being concerned,

> The question was whether or not the County would potentially revert [to traditional processes and behavior]. Here is a really big project, approximately 180,000 square feet of school on a fast track process. What would happen if they couldn't keep up?

Great Outcomes

It turned out to be a different experience altogether. Fred Knapp believed the County was open to change,

> I believe this conversation has been going on for the last couple years, especially emerging from the recession. [The County] needed to reinvent themselves and be more of an asset to the community rather than a burden—and they accomplished it with flying colors. They have been open to suggestions, and they have been open to the industry a lot more.

For their part, Fred Knapp and his team, including subcontractors, understood that they would have a part to play in making the relationship a success. They realized, early on, the benefits of having a single inspector. What an improvement it was not having to worry about new inspectors arriving on site each day that might see something different or have a different interpretation on a second inspection! The construction team wanted to maximize that advantage by making every inspection count and not wasting this inspector's time. Admittedly, Fred notes, there were some initial bumps in the process and relationship,

> This project is a big project, on fifty-seven acres with eight buildings. Sometimes we will call for multiple inspections, and no matter how many eyes I put on it, every once in a while, they will miss something. Early on, if we missed something, that might put a sour taste in the inspector's mouth. Next time he'd be thinking, are they going to miss it again. So, we had to build a rapport with that inspector and make sure that we did everything—we had to make sure we met or exceeded his expectations. The inspector didn't cause the bumps, they were caused by us being insufficiently ready when we called for our inspection. So, once we learned to prepare for those expectations of the inspector, it's been seamless.

Swaim also observed a willingness on the part of the inspector and plan reviewers to build rapport,

> They want to be part of the team culture and have some fun as opposed to always being treated as the "Code Guys." I think they loved it and the feedback I've gotten from some of the people in Pima County is that they're tweaking some of their internal environment to be able to respond more to the type of culture that we've been trying to create.

The benefits of this reversal of the original role reversal were manifest. Fred Knapp commented,

> We saw the opportunity to be able to accelerate the schedule. That seems to be a speed bump we run into on traditional projects. We have to wait for inspections, and then if there is an issue, there is reinspection and then this and that and the other. But with Pima County, we do a Skype and they can say, "You have your spacing too long on the rebar," and then we can say, "I'll call you back in an hour and have it fixed." Then they will Skype them again and sign it off. In the past, they would physically come out, inspect, leave, and then you have to reschedule an inspection, so the project would lose at least a whole day if not two [...] The other beautiful part about Skype [and the dedicated inspector] is you start to understand the personality of the inspector. We know their expectations. We built rapport and we built trust.

According to Fred Knapp, understanding these expectations also helps avoid rework. The entire team, including subcontractors, reassured about the consistency in inspections, tried harder to get the work done correctly the first time. There was adequate time to get it right the first time since they didn't have to worry about wasted repeated inspections created through adversarial relationships.

The innovations experienced in this reversed and rearranged relationship with the Pima County Building Department have been instructive to all the stakeholders involved. Fred and Phil felt all team members, including the inspector, had revised perspectives and attitudes about the roles of their teammates. Pima County continues to innovate in its processes at the building department, implementing and improving electronic submission and review services, dramatically improving review turnarounds. Lessons learned from Mica Mountain are being rolled into other Pima projects to improve inspector/contractor relationships and to build value for owners.

When Fred Knapp first heard the idea from Phil about asking Pima County to join the team, his first reaction was, "Why would we ever want to do that? It was such a relief to get them out of the process, and, now, you want to let the fox back into the henhouse?" Now, with the delivery date for Mica Mountain High School barreling down at the team, Fred is more than delighted to have this particular fox as a teammate.

Research

7

In 1967, the minimum wage for major league baseball players was the same as in 1947, $6,000 per year despite continuous inflation in the postwar era. The baseball owners, appropriately referred to as Lords in John Helyar's *The Lords of the Realm: The Real History of Baseball*, did not think of players as partners in their businesses or even as workers. The Lords thought of these men as mere players whom the Lords condescended to allow to play a game for a living and who often needed coddling, protecting, and discipline. The Lords and their general managers tricked and abused the players regarding salary negotiations and misrepresented player benefits, combining pension funds with operating funds. The Lords even refused to allow arbitration. All player appeals went straight to the commissioner of baseball, a man that the Lords, the owners, hired and fired.[1]

When Marvin Miller, director of the newly created Players Association, arrived in the mid-sixties, the players were suspicious of the idea of a union and skeptical that organized negotiations could benefit them. They had been the victims of high-handed bosses, the owners, for a long time and had no desire to get involved with a labor boss who would also be ordering them around. But Marvin Miller was a very different kind of leader. He was a teacher, a motivator, a supporter, and, oddly, an economist.

Miller understood the value of knowledge. To negotiate the first Basic Agreement with the Lords, including minimum salary, Miller needed lots more information than the Lords were willing to share. To solve this problem of lack of salary transparency, Miller enlisted the players.

> Miller asked each player to write his salary on a slip of paper and give it to his team's player rep. The player reps would collect and send them

to Miller. It would all be anonymous but quite accurate. Miller quickly disseminated the data to the players. The average salary was $19,000. About 33 percent of all players were at $10,000 or less. More than 40 percent were paid $12,000 or less. Though only 7 percent of players were at the base $6,000, raising the minimum would lift many other players' salaries as well.[2]

The results profoundly shocked the players. Some players were outraged. It was clear from the data the Lords had been misrepresenting more than the information regarding the lowest paid players. They had been lying to the highest-paid players as well. This experience drove home the idea to the players that they had been negotiating with far less knowledge than the owners. It was time to get organized, get educated, and get more data.

The Knowledge We Need

Built environment research examines the resource, health, economic, policy, and cultural impacts (and opportunities) related to the human constructed environment. This research also examines the processes, performance, and consequences of building this environment. – University of Arizona, RESTRUCT[3]

Shaping and constructing our built environment is serious business because almost nothing impacts our health, economy, culture, well-being, resources, or climate so much as the built environment. The waste, noise, and disruption generated by our antiquated construction processes stress our oceans, air, cities, and neighborhoods beyond reason. Hurricanes, wildfires, and tornados ravage our communities with increasing frequency, yet the innovation opportunities presented by our infrastructure, transportation systems, protective structures, and temporary construction to address these challenges go unrealized. Informal housing has emerged as the only recourse for millions of people in economically challenged urban areas. These self-built homes sprout in days or weeks without water or sanitary infrastructure, often because municipalities throughout the world have no understanding of how to begin to affordably address basic housing needs at scale. Today, in 2020, hospital leaders, higher education presidents, real estate developers, directors of long-term care facilities, and small business owners are all desperately trying to understand how the built environment undermines or supports their efforts during the current pandemic, but the

Commercial and residential buildings in the United States consume about ····· **70%** of our electricity ····· and **40%** of our raw materials

40% of total energy

36% of all global energy consumed is related to building construction and operations.

Buildings account for nearly **40%** of energy-related CO_2 emissions globally.

136 **MILLION TONS** of building-related construction and demolition debris are generated in the U.S. in a single year.

This waste ends up in landfills all over the globe constituting

29% of U.S. landfills   **44%** of U.K. and Austrlian landfills   **35%** of global landfills

Figure 7.1 Data demonstrates the impact of the built environment

research and literature available to answer questions about air flow, surface transmission, or even the effectiveness of filters is far too thin to provide a clear path of action.

Decision-makers in governments and boardrooms throughout the world need more and better knowledge regarding the built environment to make wiser choices. Yet, the amount of built environment research completed and shared in a holistic, credible fashion to solve the built environment's challenges is almost minuscule compared to other research areas. We cannot answer with confidence when we ask, as decision-makers, why one delivery process is better than another, how housing design can impact a community's culture, or how mechanical systems can effectively minimize COVID-19 virus air transmission. The available knowledge resources are not robust enough.

Reprising the Hard Truth about Efficiencies and Productivity

In Chapter 1, I regaled you with data about the enormous amount of money lost related to our industries' unpredictability and inefficiency. The ramifications are manifest, as noted by McKinsey researchers in 2016, stating, "The construction industry is ripe for disruption," not only because of these inefficiencies and the lack of predictability but because "financial returns for contractors are often relatively low—and volatile."[4] This significant understatement is echoed in Mark Farmer's report on the condition of the UK construction model in 2016, aptly subtitled "Modernize or Die." Farmer succinctly documents the troubles of the construction industry to include "low productivity," "low predictability," "leadership fragmentation," and "lack of collaboration improvement culture," among others. Farmer goes on to document extensive lost opportunities for the industry and the resulting vulnerabilities to individual players as well as low profitability. "The industry has evolved a 'survivalist' shape, structure and set of commercial behaviours in reaction to the environment in which it operates. That environment is fundamentally characterised by low capital reserves and high demand cyclicality."[5]

Although Farmer observes in his report that there is a need to increase R&D to address efficiency and productivity, it is another 2017 McKinsey report that, perhaps, indirectly brings this need into full focus. In *Improving Construction Productivity*, McKinsey consultants Barbosa, Mischke, and Parsons make seven specific recommendations to improve the construction industry as follows:

1   Reshape regulation and raise transparency
2   Rewire the contractual framework
3   Rethink design and engineering processes
4   Improve procurement and supply-chain management
5   Improve on-site execution
6   Infuse digital technology, new materials, and advanced automation
7   Reskill the workforce[6]

It must be noted, each of these recommendations to improve the design and construction delivery process requires knowledge that does not yet exist, knowledge developed through structured, credible, methodical research.

The hard truth is we are leaving too much money and capacity on the table. Twice in this book, I quoted from *Building the Future*, which suggested 75 percent of construction activities may add no value.[7] In "Reinventing Construction," McKinsey researchers expressed the issue of misaligned resources more pointedly,

> Today, around $10 trillion a year is being spent on the buildings, infrastructure, and industrial installations that are the backbone of the global economy, and demand is rising. By 2025, that amount is projected to total $14 trillion. However, the industry could produce more for this investment if productivity were higher, leading to a fundamental improvement in the world's infrastructure and the quality of life of citizens.[8]

Impact on Resources and the Environment

If you have any doubts about the impact of the built environment on our resources, consider these facts. Edmondson and Reynolds wrote in *Building the Future*, "Commercial and residential buildings in the United States consume about 40 percent of total energy, including 70 percent of our electricity, and 40 percent of our raw materials."[9] This startling summary is supported by a US Green Building Council (USGBC) statement reporting that buildings account for 70 percent of all US resource consumption as well as 12.2 percent of all potable water, 39 percent of primary energy, and 40 percent of all raw materials globally.[10] The Global Alliance for Buildings and Construction reported 36 percent of all global energy consumed is related to building construction and operations.[11]

The facts regarding the effects the built environment has on our natural environment are even more chilling. The Global Alliance for Buildings and Construction also reported buildings account for nearly 40 percent of energy-related CO2 emissions globally,[12] consistent with reports allocating 45 percent of the United Kingdom's total carbon emissions to buildings. Commercial structures comprise 27 percent of that amount.[13]

Another environmental concern is construction and demolition waste. Some estimates assert "as much as 30% of the total weight of building materials delivered to a building site" leaves that site as construction waste.[14] The USGBC reported, "136 million tons of building-related construction and demolition debris is generated in the U.S. in a single year (By way of comparison, the U.S. creates 209.7 million tons of municipal solid waste per year)."[15] This waste ends up in landfills all over the globe, constituting 29 percent of US landfills, 44 percent in the United Kingdom and Australia, and 35 percent globally.[16] Construction Dive reported in 2018 construction waste may nearly double to 2.2 billion tons worldwide by 2025.[17] This waste not only affects the environment but adds to the already unreasonable cost of construction.

Growth and Health

While we struggle with inefficiencies with our process and our product, the world continues to grow, and its infrastructure continues to age. Cities throughout the world have experienced unprecedented growth, with cities of millions appearing in just a few years. *Architecture 2030*, prior to the pandemic, anticipated two-thirds of the world's population will live in urban areas by 2060, equivalent to nearly 6.6 billion people living in cities. This level of growth and migration would require an addition of "2.48 trillion square feet (230 billion m2) of new floor area to the global building stock."[18] Until the global pandemic, construction revenue growth was also on an unprecedented trajectory. McKinsey had projected in 2017 a $14 trillion worldwide industry by 2025,[19] and Global Construction Perspectives had forecast a $15 trillion target.[20]

Regardless of whether revenue projections hold up during the post-pandemic economic cycle, the world population is still growing. Although population growth has slowed in the last half century, the world will add at least 3 billion people to the current 7.7 billion by the end of our century.[21] In 2018, UN-Habitat metrics estimated

> 1.6 billion people live in inadequate housing globally, of which 1 billion live in slums and informal settlements. This means that about one in four people in cities live in conditions that harm their health,

safety, prosperity and opportunities. Lack of access to basic services is a common constraint in informal settlements and slums: worldwide 2.4 billion people live without improved sanitation and 2 billion are affected by water stress.[22]

Most expected urban growth will be within developing countries where urban populations will double over thirty years. Parul Agarwala and Syed Usman Javaid wrote in 2013, "3 billion people will need housing and basic infrastructure. Already, 70% of existing housing in developing countries is built informally without appropriate structural standards."[23]

While growth pressures urban areas and housing, infrastructure is crumbling. The American Society of Civil Engineers (ASCE) has given America's infrastructure a "D+" related to current conditions and the need for capital reinvestment. Wastewater, drinking water, roads, levees, bridges, energy, dams, railways, and transit are included in this assessment along with aviation, parks, schools, solid waste, hazardous waste, and inland waterways. Much of this sobering data directly relates to the built environment. For example, the ASCE estimates the United States needs to spend $4.5 trillion by 2025 to fix the country's roads, bridges, dams, and other infrastructure. The association estimates there were over 56,000 structurally deficient bridges in 2016 and 188 million daily trips across those structurally deficient bridges. "One out of every five miles of highway pavement is in poor condition and our roads have a significant and increasing backlog of rehabilitation needs." For drinking water infrastructure, which received a "D" all by itself, the Report Card noted that many of the one million miles of pipes delivering drinking water in the United States were laid in the "early to mid-20th century with a lifespan of 75 to 100 years." These pipes break approximately 240,000 times per year, accounting for a loss of two trillion gallons of treated water. The ASCE reports hopefully, "The quality of drinking water in the United States remains high, but legacy and emerging contaminants continue to require close attention."[24]

Growth, infrastructure, and related health challenges are products of both the neglect of our industries' professionals and a lack of critical knowledge. How do we solve these problems in a fully informed and effective manner?

Barriers to Knowledge-Building

For years, research on the built environment has been either neglected entirely, has been approached in silos, has manifested as undisciplined experimentation mistaken for credible research, or has been perceived as just too overwhelmingly complicated. It's time to look at how research on the built environment can be addressed differently, so we solve problems in a way that is fully informed, impacts decisions, changes behavior, and shapes policy.

Lack of Investment

In 2016, McKinsey researchers were candid, "R&D spending in construction runs well behind that of other industries: less than 1 percent of revenues, versus 3.5 to 4.5 percent for the auto and aerospace sectors."[25] This investment problem not only exists on the industry level. It also exists on the federal level. The United States has research funding agencies for health, science, agriculture, parks, oceans, defense, space, education, environment, disabilities and justice, but has no agency dedicated to building the constructed environment safely, effectively, economically, and in a manner that protects the welfare and well-being of its citizens. A USGBC report, published in 2007, documented "research related to high-performance green building practices and technologies" was equal to approximately 0.2 percent of all federally funded research, "just 0.02% of the estimated value of annual U.S. building construction."[26]

There are many reasons for this lack of investment by firms. The fragmented industry is part of the problem. As Edmondson and Reynolds noted, "Individual construction firms have little incentive and few resources to collect, integrate, or use collective wisdom."[27] Thomsen and Sanders argue, "Significant change typically requires significant investment, but construction industry organizations are reluctant to make large R&D investments because there is no direct reward: these investments are unlikely to produce returns commensurate with the cost of the investment."[28] This assessment is accurate, but as I suggested in *The Owner's Dilemma*, many of our industry professionals are not educated in the processes or the benefits of systematic research. "Architects and construction managers are simply not taught how to collect a foundation of knowledge efficiently, build on it, and then pass it along to others for further improvement."[29] Without an experience that teaches our young professionals the value of building credible knowledge, there is no reason to seek it.

Further complicating the industry research landscape is the common practice of knowledge hoarding. When firms or construction companies complete research projects, they will most often hoard the knowledge outcomes. These firms consider the new information a competitive advantage and do not leverage it with the knowledge of others, potentially redoubling its benefits. Therefore, firms rarely get the payback expected from the R&D investment. The many benefits of sharing knowledge are a hard lesson our industries still must learn.

The federal government is another issue altogether. As with most federal funding conversations, research funding is a lobbying exercise. The design and construction industries have never told their research story, perhaps because it was not well understood. Most lobbying funds have been spent on narrow issues at state levels focused on protecting professions or project access. Although there has been interest in energy and transportation for some time, the federal government is just beginning to realize that research informs building codes, saving money during natural disasters.

For example, C2SE reported,

> one of the most beneficial approaches to improve the resilience of buildings is to widely adopt the most current model building codes. The study found that building to the current common model building code standards as opposed to the model codes from the 1990s returns an average of $11 for every dollar (11:1) invested pre-disaster.

Similar benefits were reported after extensive research and testing improved building codes after Hurricane Andrew in 1992,

> A review of ten years of insurance loss data showed that the adoption of this stronger building code reduced the number of claims and the total value of claims was 72 percent less for the buildings built since 2000. In that ten-year period, Florida received $3.50 in benefit (due to the lower number and value of insurance claims) for every $1.00 of additional cost associated with implementing the building code.[30]

The federal government has recently become more aware that investing in resilience research and crisis management research in the built environment can pay off. It is time to communicate how built environment research can also support the economy, health, and welfare of our communities, as well as conserve resources.

Silos

While we are dealing with facts, we must recognize that much of the research produced in the built environment is completed in silos by the traditional disciplines. Other academic disciplines actively create some related research, and some research is multidisciplinary. However, most research is mono-disciplinary: mechanical engineering, structural engineering, civil, engineering, construction management, architecture, planning, and so on. Our most creative and inclusive discipline may be planning, known for including social scientists, traffic engineers, and many others in their studies.

It is most comfortable to stay in each of our professional lanes, but that has specific consequences. A narrow focus of a significant problem such as "Net Zero Water," for example, might only address a technical question rather than the behavior, policy, systems, business, or even formal (architectural) issues. To be implemented effectively, all these issues must be addressed as they have been in the project developed by a team at the University of Arizona. See Figure 7.2, developed by Courtney Crosson with contributing ideas from the University of Arizona Net Zero Urban Water Coordinating Committee, Dominic Boccelli, Jennifer Duan, Thomas Meixner, and Christopher Scott, illustrating the integration of disciplines and challenges leading to sustainable solutions.

Lack of transdisciplinary collaboration may also mean that the technical solution solves the problems incorrectly because it misses vital constraints. The built environment is a complicated place, and profound change will require developing critical, credible research across boundaries.

Undisciplined Experimentation

In *The Owner's Dilemma*, I wrote a naive chapter on research, specifically on scientific method.[31] My heart and intentions were in the right place; however, I did not yet know what I did not know about research. For me, the last decade has been an incredible adventure in engaging in, writing about, and building research. Most of all, I have learned a great deal about scholarship and the broad spectrum of definitions and focus of scholarship. Specifically, when engaging in research, I have learned there are many approaches to methods. However, research design, disciplined methodology, and peer review are all critical elements to credible, rigorous knowledge creation, an activity quite apart from undisciplined experimentation.

I once attended a lecture where an architect proudly described his experimentations inspired by the digital obliteration patterns printed on the inside of envelopes. Artistically, it was personally stimulating to him. However, this experimentation was not research. It may be his personal scholarship, but it is not methods-based research. It is essential to understand

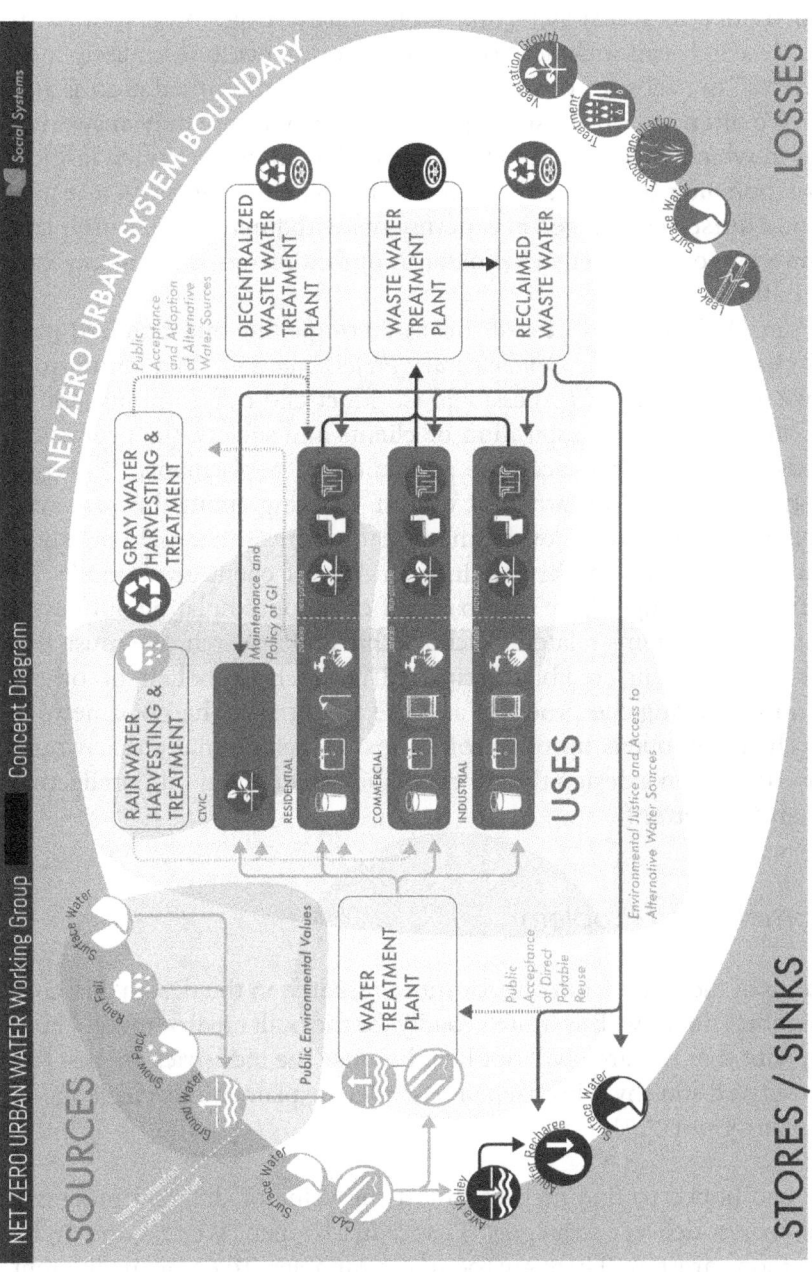

Figure 7.2 Research Diagram by Courtney Crosson with contributing ideas from the University of Arizona Net Zero Urban Water Coordinating Committee, Dominic Boccelli, Jennifer Duan, Thomas Meixner, and Christopher Scott

the difference. Perhaps because architects think of themselves as artists and craftspersons as well as professionals, and probably because architects are not often trained as researchers, we often confuse experimentation with research. Research can include structured experiments, but undisciplined "let's try this" or "let's go down the rabbit hole" experimentation is not research. We often, as a profession, spend most of our time only answering questions in which we are personally interested in an undisciplined fashion instead of pursuing questions that need a disciplined answer. As a result, we may find ourselves engaged in experiments with limited application and limited interest to any other stakeholder in our industries, a luxury we can no longer afford.

Benjamin Derbyshire expressed this as a compelling and required professional change in *Defining Contemporary Professionalism*, "In addition to a newly articulated Code of Professional Conduct and our commitment to research on the needs and aspiration of clients and society, the profession now needs to work on the accurate prediction of performance."[32] Peggy Deamer has an even more practical way of thinking about this research, "we emphasise research and development that produces materials, products and techniques that are marketable independent of client invitation."[33] As an industry, we no longer have the luxury of only contemplating our navels and following our whims related to scholarship and research. We must also endeavor to engage in credible disciplined research on behalf of others focused on improving the products and the industries. The good news is that this challenge opens up our professions to include many that would never have thrived in a design theory class. We can broaden our intellectual diversity and be better for it.

It's a Complicated Problem

Even after a decade of advocating for more research in the design and construction industries as well as more broadly for the built environment, I hear the complaint that research will not help because the industries are just too complicated and squishy. Thomsen and Sanders explained this challenge in *Program Management 2.0*,

> We had hoped to find metrics to test the value of different processes and project delivery strategies. That didn't happen. We couldn't find a common yardstick. There are too many variables. To compare different processes to determine the best, you must hold all the variables but one constant. We can't do that with design and construction. Good

comparative analysis requires a researcher to hold all variables but one constant to measure the effect of that variable [...] We can't build two projects at the same time, on the same site, with the same people and simply change one thing [...] Sometimes there is overwhelming evidence of the value of a good practice [...] But even then, there will be skeptics.[34]

Thomsen and Sanders go on to discuss the problems of dealing with the definitions of quality in design and construction, calling it "a personal, fuzzy definition" frequently changing and hard to pin down.[35] All of these characteristics make research in the built environment a challenge, no doubt, but far from impossible.

This conundrum is where the overall lack of understanding of research methods in our industry causes so much confusion. There are many more research methods than the "double-blind study," which we can use to delve for answers. Imagine trying to define the essential, most beneficial characteristics of collaboration and leadership as well as which of those characteristics have the most significant impact on producing high-performance buildings. If this research project sounds complicated, it was. The research project, Renée Cheng's study "Teams Matter," studied eleven projects from the GSA American Recovery and Reinvestment Act (ARRA) with a daunting description:

ARRA projects shared a high level of complexity and risk in the following categories: cost, budget, schedule, technical design, logistics, and transactions. All projects involved renovation, and interfacing with existing building systems raised technical (and often unforeseeable) challenges. In some cases, technical complexity was increased by the integration of innovative high-performance building technologies with outdated systems. Logistics were complicated as tenants typically remained in their units during the renovation, introducing security and operational concerns.[36]

In this study, Cheng and her team operationalized thirty-seven different characteristics of context, key ingredients, and outcomes using case study construct and qualitative comparative analysis methodologies to collect and analyze data. Data included documentation, interviews, and surveys. The study provides, along with clear definitions, individual project insights, and clear analysis, an amazing "Truth Table" synthesizing the essential information regarding differences in context, and key contractual, leadership, and process characteristics related to quality of outcomes. Suddenly, the extraordinarily complex, after hours of painstakingly methodical work, is clear and understandable.[37] Cheng shows us one way to grapple with industries that seem too dense and complex to understand. We can delve into the unknowable underlying the design and construction industries and emerge with priceless gems of knowledge.

Vision for the Future

Research-informed design and processes can address all the issues articulated in this chapter. We have been flying blind for so long as professionals, it is embarrassing. More awkward is our comfort in using the perspectives, design approaches, and construction processes of past generations despite having the tools of the future at hand. **For research to serve our future and the future of the built environment, it must be transdisciplinary, robustly funded, shared, data-driven, and focused on creating an adaptive future.**

The serious business of shaping and constructing the built environment means opening ourselves to the ideas, knowledge, and experience of stakeholders and disciplines outside the usual building professions. As Indy Johar, architect and founder of Architecture 00 observed in *Defining Contemporary Professionalism,*

> Design practice is beginning to comprehend and metricise its capacity to effect change through behavioural economics, econometrics and cognitive psychology. These fields are proving the intuitive knowledge of architects on the factors, tools and design tactics that impact human decision-making, influencing behaviours and thereby empathy and culture.[38]

Understanding the decisions we make as designers cannot be made in isolation and these decisions impact well-being, economics, and resources must make us curious regarding the knowledge that other disciplines hold. More importantly, making sustainable, sticky change requires the ability to predict behavior and to understand policy. All of these aspects work together as a system, so understanding systems dynamics becomes an urgent talent.

For these reasons, the University of Arizona has, as part of its strategic plan, created the first pan-university, all-inclusive transdisciplinary approach to research in the built environment. At the time of writing, thirteen of the university's colleges are part of this initiative including public health, optical sciences, social and behavioral sciences, education, fine arts, medicine, science, agriculture, humanities, nursing, and law, along with the typical built environment colleges, engineering, and CAPLA (architecture, landscape architecture, and planning). The initiative, called RESTRUCT, has more than thirty external civic and industry partners engaging in the development of research projects and collaboratively seeking grants.[39]

The faculty participating in this initiative identified four grand challenges based on existing university strengths as follows:

- Redress Inequality / Injustice in re-envisioning the Built Environment
- Create Resilient and Efficient Urban and Rural Systems
- Design for Optimal Heath
- Enable Innovation through Better Decision-Making & Data Analysis[40]

As part of this initiative, the UArizona teams have recognized the vast resources of existing data related to the built environment within municipalities, institutions, and projects. Cities may measure traffic, pedestrians, heat island effects, cycling accidents, even real estate characteristics, yet the associated data may be defined and measured differently. Often these data are not accessible to researchers at all. Creating leveled, defined warehouses of built environment data, available to all researchers, is a worthy goal in which many universities may be interested. The University of Arizona is actively working on these opportunities. During the pandemic, teams from RESTRUCT have worked on mobile testing facilities, developing processes for restarting the hospitality industry, and other projects for supporting facilities reentry on university campuses related to airflow, queuing, anti-microbial surfaces, and documenting higher-risk environments.

All of these projects are most effectively addressed through transdisciplinary teams. The mobile testing facility team had twenty members, including five members external to the university, providing design, construction, and public health expertise. Internal members provided expertise in building performance, health outcomes in the built environment, team building, computation and decision-making, smart product design, airflow and temperature control, modular construction, queuing optimization, stress testing in the built environment, sensors and air quality, biomedical innovation, infectious diseases, health disparities, epidemiology, public sector collaborations, and risk assessment. This team and their convergent approach created a winning proposal addressing not only structural issues but health and process issues in only one week.

Transdisciplinary research presents a new future for architects, engineers, planners, and construction managers. The transdisciplinary approach to research offers excellent potential for finding answers that are practical and actionable. However, we must have a research culture open to this activity, including knowledge sharing across academia and industry. Again, Indy Johar advocates as I do,

Unleash the collective knowledge, intelligence, innovation and belonging of all. Enhance a culture around data and evidence—hypothesis-driven architecture—by establishing new demand using public procurement requirements and PhD-led research. Create a new data commons for all architects and their design data/information. Establish a viable open source file standard for design data. Reset the performance metrics of architecture, creating a built environment focused on evidence-based design and the performative effect and on the social and environmental outcomes. Advocate new schools of architecture delivering a curriculum focused on a systems approach and design driven by analysis, evidence and benefit.[41]

How will we share research once it is created? There are a limited number of academic journals that will help a bit. However, these journals aren't focused on the transdisciplinary research and outward-facing research we need to share so desperately. These journals are also not collecting or sharing industry research, a huge gap that is harder to overcome than it looks. London-based architect Robin Nicholson described this as one of the challenges of various RIBA projects,

Faster dissemination of research—Once the penny dropped that practitioners were more interested in being aware of new issues than studying them in detail, eight of our institutions pulled together and we were confident that a new multi-disciplinary e-journal was imminent. But this much-needed easy-win has proved difficult to deliver.[42]

How do you make an industry-based, multidisciplinary journal work? For credibility, it must be peer reviewed. How will that work? More importantly, will the industry begin to share essential knowledge worth publishing?

Adaptive Design and Construction

COVID-19 has provided the latest evidence the built environment is integrated with almost every challenge we face related to our health, economy, culture, resources, and environment. Multiple questions have emerged related to how we can live and conduct business while remaining safe from contagion, but also regarding the impact of our "new normal" on real estate, housing, high-rise buildings, mechanical systems, and energy consumption. This crisis expands the need for inquiry and compels us to seek more adaptive and resilient approaches to the built environment. Through improved access to data, enhanced modeling, and removal of critical information barriers, we will ultimately improve decision-making as we address knowledge gaps related to the processes, performance, and consequences of constructing the world around us.

Those of us in the design and construction industries are aware we have essentially been designing, building, even thinking about our constructed environment in the same manner for centuries. As we design and build, we generally think of constructed elements as permanent instead of resilient structures capable of adapting to time, needs, climate, and crises. In 1994, Stewart Brand wrote a book titled *How Buildings Learn: What Happens after They're Built*, which recognized that buildings begin adapting the day they start being used, unlike the precious visions of their architects.[43] I learned from this book that all buildings, homes, offices, and institutions, change at different speeds in different layers. The structure of the building changes less often than the interior walls. The mechanical system changes more often than the exterior skin, for example. To comprehensively address the increasing challenges of our health, society, and economy affected by the built environment, universities, industry, and civic leaders must embrace a fully integrated and adaptive approach to problem-solving and decision-making.

Given what we know today about our resources, climate, culture, technology, and health challenges, shouldn't what we build be ready to change more easily? Shouldn't we rethink how our built environment could or should respond to our needs whether we are isolated or gathered together?

We have no idea what will happen in the next six months, year, or five years. The market, which has been on a roller-coaster ride over the last several, seems equally unequipped to predict the future. However, research, carefully planned and methodically crafted, can help us create a more adaptable and resilient built environment so that our future response to crises can be more predictable.

Notes

1 John Helyar, *The Lords of the Realm* (New York: Ballantine Books, 1994).

2 Ibid., 29.

3 "RESTRUCT Symposium Opening Presentation," RESTRUCT, December 12, 2019, RESTRUCT.arizona.edu

4 Rajat Agarwal, Shankar Chandrasekaran, and Murkund Sridhar, "Imagining Construction's Digital Future," McKinsey Productivity Sciences Center,

Singapore, Capital Projects and Infrastructure, June 2016, www.mckinsey.com / ~ / media / McKinsey / Industries / Capital%20Projects%20and%20Infrastructure / Our%20Insights / Imagining%20constructions%20digital%20future / Imagining-constructions-digital-future.ashx, 2.

5 Mark Farmer, "The Farmer Review of the UK Construction Labour Model: Modernise or Die," Construction Leadership Council, October 2016, 8.

6 Filipe Barbosa, Jan Mischke, and Matthew Parsons, "Improving Construction Productivity: Voices," McKinsey Global Infrastructure Initiative, June 2017, 3–5.

7 Amy C. Edmondson and Susan Salter Reynolds, *Building the Future: Big Teaming for Audacious Innovation* (Oakland, CA: BK / Berrett-Koehler, 2016) 107.

8 Filipe Barbosa, Jonathan Woetzel, Jan Mischke, Maria Joao Ribeirinho, Mukund Sridhar, Matthew Parsons, Nick Bertram, and Stephanie Brown, "Reinventing Construction: A Route to Higher Productivity", McKinsey Global Institute, February 2017, www.mckinsey.com / ~ / media / McKinsey / Industries / Capital%20Projects%20and%20Infrastructure / Our%20Insights / Reinventing%20construction%20through%20a%20productivity%20revolution / MGI-Reinventing-Construction-Executive-summary.ashx, 8.

9 Amy C. Edmondson and Susan Salter Reynolds, *Building the Future: Big Teaming for Audacious Innovation* (Oakland, CA: BK / Berrett-Koehler, 2016), 106.

10 USGBC, "Statement of the U.S. Green Building Council Before the Senate Committee on Environment and Public Works on Green Buildings: Benefits to Health, the Environment and the Bottom Line," May 15, 2007, Washington, DC, 1.

11 Brian Dean, John Dulac, and Ian Hamilton, *2018 Global Status Report: Towards a Zero-Emission, Efficient and Resilient Buildings and Construction Sector*, Global Alliance for Buildings and Construction, International Energy Agency, United Nations Environment Programme, 2018, https:// globalabc.org / sites / default / files / 2020-03 / 2018_GlobalAB_%20Global_Status%20_Report%20_English.PDF, 9.

12 Ibid.

13 "Carbon Dioxide in Construction," Designing Buildings Wiki, June 9, 2020, www.designingbuildings.co.uk / wiki / Carbon_dioxide_in_construction; site references the Technology Strategy Board (TSB) Report on UK Carbon Emissions. The TSB changed its name to Innovate UK in 2014.

14 Mahamed Osami, "Construction Waste," *Waste*, 2011, www.sciencedirect.com / topics / earth-and-planetary-sciences / construction-waste, 1.

15 USGBC, "Statement of the U.S. Green Building Council Before the Senate Committee on Environment and Public Works on Green Buildings: Benefits to Health, the Environment and the Bottom Line," May 15, 2007, Washington, DC, 1.

16 Saheed O. Ajayi, Lukumon O. Oyedele, Olugbenga O. Akinade, Muhammad Bilal, Hakeem A. Owolabi, Hafiz A. Alaka, and Kabir O. Kadiri, "Reducing Waste to Landfill: A Need for Cultural Change in the UK Construction Industry," *Journal of Building Engineering*, 5 (March 2016): 185.

17 Kim Slowey, "Report: Global Construction Waste Will Almost Double by 2025," *Construction Dive*, June 5, 2020, www.constructiondive.com/news/report-global-construction-waste-will-almost-double-by-2025/518874/.

18 "New Buildings Operational Emissions," Architecture 2030, June 3, 2020, https://architecture2030.org/new-buildings-operations/.

19 Filipe Barbosa, Jonathan Woetzel, Jan Mischke, Maria Joao Ribeirinho, Mukund Sridhar, Matthew Parsons, Nick Bertram, and Stephanie Brown, "Reinventing Construction: A Route to Higher Productivity", McKinsey Global Institute, February 2017, www.mckinsey.com/~/media/McKinsey/Industries/Capital%20Projects%20and%20Infrastructure/Our%20Insights/Reinventing%20construction%20through%20a%20productivity%20revolution/MGI-Reinventing-Construction-Executive-summary.ashx, 8.

20 Global Construction Perspectives Data as documented in the following paper: Saheed O. Ajayi, Lukumon O. Oyedele, Olugbenga O. Akinade, Muhammad Bilal, Hakeem A. Owolabi, Hafiz A. Alaka, and Kabir O. Kadiri, "Reducing Waste to Landfill: A Need for Cultural Change in the UK Construction Industry," *Journal of Building Engineering*, 5 (March 2016): 185. Original current data can be purchased at www.globalconstruction2025.com.

21 Max Roser, Our World in Data, June 9, 2020, https://ourworldindata.org/future-population-growth#global-population-growth.

22 United Nations Human Settlements Programme UN-Habitat, "Indicator 11.1.1: Proportion of urban population living in slums, informal settlements or inadequate housing, Fundaments of Urbanization, Evidence Base for Policy Making." 2016 (Updated February 14, 2018, https://unstats.un.org/sdgs/metadata/files/Metadata-11-01-01.pdf.

23 Parul Agarwala, and Syed Usman Javaid, "Is Upgrading Informal Housing a Step in the Right Direction?" World Bank Blogs, World Bank.org, July 9, 2013, https://blogs.worldbank.org/endpovertyinsouthasia/upgrading-informal-housing-step-right-direction.

24 "2017 Infrastructure Report Card", ASCE, June 6, 2020, www.infrastructurereportcard.org.

25 Rajat Agarwal, Shankar Chandrasekaran, and Murkund Sridhar, "Imagining Construction's Digital Future," McKinsey Productivity Sciences Center, Singapore, Capital Projects and Infrastructure, June 2016, www.mckinsey.com/~/media/McKinsey/Industries/Capital%20Projects%20and%20Infrastructure/Our%20Insights/Imagining%20constructions%20digital%20future/Imagining-constructions-digital-future.ashx, 2.

26 USGBC, "Statement of the U.S. Green Building Council Before the Senate Committee on Environment and Public Works on Green Buildings: Benefits to Health, The Environment and the Bottom Line," May 15, 2007, Washington, DC, 8.

27 Amy C. Edmondson, and Susan Salter Reynolds, *Building the Future: Big Teaming for Audacious Innovation* (Oakland, CA: BK/Berrett-Koehler, 2016), 107.

28 Chuck Thomsen and Sid Sanders, *Program Management 2.0: Concepts and Strategies for Managing Building Programs* (rev. ed.) (McLean, VA: Construction Management Association of America Foundation, 2011), 28.

29 Barbara White Bryson and Canan Yetmen, *The Owner's Dilemma: Driving Success and Innovation in the Design and Construction Industry* (Atlanta: Ostberg Library of Design Management, Greenway Communications, 2010), 204–205.

30 C2ES Center for Climate and Energy Solutions, *Investing in Resilience*, November 2019, www.c2es.org/document/investing-in-resilience/, 2.

31 Barbara White Bryson and Canan Yetmen, *The Owner's Dilemma: Driving Success and Innovation in the Design and Construction Industry* (Atlanta: Ostberg Library of Design Management, Greenway Communications, 2010), 205–207.

32 Alan Jones and Rob Hyde, eds., *Defining Contemporary Professionalism: for Architects in Practice and Education* (London: RIBA Publishing, 2019), 70.

33 Ibid., 66.

34 Chuck Thomsen and Sid Sanders, *Program Management 2.0: Concepts and Strategies for Managing Building Programs* (rev. ed.) (McLean, VA: CMAA, 2011), 76.

35 Ibid., 77.

36 Renée Cheng, "Teams Matter: Lessons from ARRA, GSA Region 5 and the American Recovery and Reinvestment Act," School of Architecture, University of Minnesota, GSA Region 5, 4240 Architecture, May 2015, 3.

37 Ibid., 10.

38 Alan Jones and Rob Hyde, eds., *Defining Contemporary Professionalism: For Architects in Practice and Education* (London: RIBA Publishing, 2019), 157.

39 "RESTRUCT," December 12, 2019, RESTRUCT.arizona.edu

40 "RESTRUCT Grand Challenges, https://restruct.arizona.edu/grand-challenges.

41 Alan Jones and Rob Hyde, eds., *Defining Contemporary Professionalism: for Architects in Practice and Education* (London: RIBA Publishing, 2019), 158.

42 Ibid., 184.

43 Stuart Brand, *How Buildings Learn: What Happens after Their Built* (New York: Viking, 1994).

Chapter Supplement—7
Asking Hard Questions

The Beginning of a Beautiful Friendship

Michael McCormick, associate vice president for asset management and university architect at the University of Washington, was exuberant about the announcement. The university had just announced the selection of the new dean of the College of Built Environments as Renée Cheng, the former associate dean of research at the University of Minnesota. Mike was familiar with Renée's research work on teams, leadership, and collaboration and had met her in 2018 at a workshop led by Howard Ashcroft on integrated project delivery (IPD) supported by a Pankow Grant. This work led to the publication of *Integrated Project Delivery: An Action Guide for Leaders*.[1] Mike had also led collaborative IPD projects during his tenure at Ivy League institution Brown University with great success. Now at the University of Washington (UW), affectionately called "UDub" by locals, Mike was combining IPD with progressive design-build in a much different environment, a major public university. When Mike arrived at UW, the university had a track record of projects going slightly over budget—not by much—only an average of 5 percent. However, over the entire building program, 5 percent totaled approximately $100 million, a significant sum that could have built an entirely new building if recaptured.

Mike's goal upon arriving at UW was to create a collaborative project delivery process that arrested overruns and built additional value for the university. That goal was a tall order, especially since the State of Washington did not allow IPD contracts on public projects. However, a thoughtful recrafting of the progressive design-build contract allowed the university to incorporate IPD processes and culture into their projects, even incentive structures. This was a new process that had a real chance of succeeding, but Mike knew he had to prove to administrators, now and in the future, that the process truly delivered value. It was essential to measure and test the delivery methods and contracts so that claims of success were credible. To that end, Mike was already working with two researchers at the College of Built Environments, Carrie Sturts Dossick and Laura Osborne, who frequently partnered with Cheng, on IPD research.

Mike was sure Renée Cheng would be a significant ally and partner in his efforts to enhance the UW delivery system, providing research to support decisions. He saw her arrival as an extraordinary opportunity, so remarkable, he sent her an e-mail to introduce his ideas even before she could pack her bags for the move. Renée Cheng was also delighted. She was looking forward to coming to the stunning UW campus not only to lead the College of Built Environments, but to collaborate on research projects with Mike McCormick, his remarkable teams, and her long-time UW research partners. Renée's research work on teams, especially the work on the GSA American Recovery and Reinvestment Act,[2,3] is nationally recognized as groundbreaking and, possibly, the first credible work on collaboration and leadership for the design and construction industries widely transferable to almost any project. Working with a like-minded practitioner open to the benefits provided by research informing processes and team building, accelerated potential opportunities for the university and the college. Renée noted the similarities in their mindsets, "Mike and I are starting from such a strong base of shared belief—integration as a way to move forward, and culture change among teams as being important for improved outcomes."

Alignment

When Renee arrived at UW, she jumped immediately into opportunities to build bridges with the facilities administration presented by Mike and his boss, Lou Cariello. With Lou, Renée supported work with other deans on buildings and landscapes at UW. With Mike, Renee reached out to various groups, internally and externally, to create shared language around integration, culture, and collaboration and to create a "baseline understanding of the research Renee had done." Concurrently, Renee identified a small amount of research funding awarded to support her research by the Lean Construction Institute and made it available for a first UW project. Mike found matching funds.

Renée saw the number of IPD projects Mike had completed at Brown and the new projects underway at UW as a significant tranche of research material to be mined to understand IPD-type projects in higher education. This is an area of study not previously explored at any length or depth. The opportunity was clear, but aligning interests and expectations was a bit more complicated. Mike and Renée admitted they were not, in the beginning, in complete accord regarding research goals or even research questions to be answered. After some discussion, they realized there were overarching questions that caught both their interest. "Is higher education unique as an owner type? If so, how? And, why has IPD been so slow to gain acceptance in higher education?" Mike queried,

> is [design and construction] inherently unique in institutions [of higher ed] and is there something that we ought to be thinking differently about as we set up our IPD projects at an institution than it would be at a hospital for example?

These were big questions that would take time, money, and a number of research projects to answer, but Renée and Mike had only a small amount of funding to start. Though it may seem risky to some, jumping in before knowing the entire scope of research was Renée's preferred strategy. Renée reflected,

> I don't tackle the very large research projects until I've completed some smaller ones that clarify the question. My typical way of working is to try to use shorter, smaller projects to better define the larger question. Then we can go with a larger set of case studies or a more in-depth methodology that would involve more analysis or other types of quantitative methods.

As Renée became more familiar with the UW design and construction program, she realized the integrated delivery programs, now called Integrated Design-Build, were very new to the university and their teams. She wondered, therefore, what a useful, informative first small project might be. As Renée reflected on the big questions regarding IPD in higher education, she wondered what the project teams were experiencing from the perspective of IPD and what is being asked of each team member? She asked, "Are the team members getting what they need from UW to be successful?"

Mike had several UW projects in various stages of design and construction, or getting close to done. Renée and her team, comprised of Dossick and Osborne, selected projects for tracking, the Health Science Education Building, Kincaid Hall Renovation, Parrington Hall Renovation, Population Health Facility, and Founders Hall. All projects were using the integrated design-build contract. Ultimately, the research project was designed as a case study, "looking at the needs of the project team and whether or not the UW, as an owner/client [...] was giving the project teams what they needed to succeed." From this small, limited study, UW would be able to gain insights into how to better serve the needs of project teams to, in turn, get better results.

As enthusiastic as Renée and Mike were about the study, it wasn't clear the project team members would be as excited about the process. The teams were busy with their work and not thrilled about taking on new responsibilities associated with this research. However, the research team wasn't asking them to produce new deliverables or to pull old documents. Renée remembered, "This was not intended to be a gigantic research project over multiple years with lots of research team members. This was meant to be a small, nimble project." Fortunately, the routine monthly report generated by the project managers provided all the documentation needed. Personal interviews provided the remaining data. "We were cautious," Renée reflected, "not to ask questions or collect data that couldn't be addressed with the amount of time that we had as a research team to analyze, or the amount of time that they had as a project team to produce."

Mike remembered each project team approached this research differently,

> The various teams are usually composed of different people from different firms. I would say that the one team that's doing two of our really high-profile projects with the same architecture firm, same consultants, same contractor, and generally same people, embraced it with open arms—because they feel like they are learning from the process and they were looking for any way possible to help accelerate that process. We have another team that took the attitude that this [research] is going to eventually be published and we don't want to look bad. So, they started to sanitize all of the interviews to the point where it wasn't really very helpful [...] That team, not surprisingly, is not the most-high functioning team. They're not very introspective—just trying to get to the end of their project.

Interestingly, the teams fully engaging in the research process were "eager to get the results, and very much wanting to hear what other teams were experiencing." The teams also seemed to understand that the effort would benefit the projects, the university, and higher education more broadly.

Outcomes and Observations

Although the research and report were not complete at the time of the interviews, several observations and indirect outcomes had already emerged. One observation reinforced Mike's past experiences,

> I would say that when the team is functioning well, the university's dysfunction [in decision-making] shows up clearly, but if the team is not functioning very well, the university's dysfunction does not show up in the same way, because the team's own dysfunction is getting in the way.

In other words, the typical university decision-making process has a hard time catching up to the high-performing project teams.

Mike reflected,

> We generally try to create a governance structure on each project that is effective at making decisions and so as long as the university is comfortable with that, it works really, really well. When it doesn't work so well is when we have a new [administrator] come in and [they aren't] as comfortable with that governance structure. They start to intervene in [a great many] decisions. You can imagine what that does to the team and to the projects. But as long as we let the governance structure work, it works amazingly well, and it can keep up with the team effectively.

Another observation arising from the research activities is the recognition that serial owners should fund built environment research. Mike outlined the current state of research in the design and construction industries, which do not traditionally carve out funding for research. However, serial owners such as universities are capable of dedicating 1 percent or a half percent of their building construction budgets to research, similar to that often dedicated to public art. These serial owners would benefit enormously from these investments. Mike remarked,

> I've given a couple talks about the power of thinking of our [projects] as a program and not just as a list of individual projects. We have some of the best researchers in the country here on campus, and we have one of the larger building programs in the country. So, for us to be a living laboratory for [built environment] research might be hugely impactful.

Renée agreed,

> At various times, the US General Services Administration (GSA) had [one] percent for research. Not every project used it, but the agency had the ability to use it if they wanted to. The GSA and the Office of High Performing Federal Buildings have sponsored many research projects, including some of my research. It's not only firms and project teams that should fund research, but funding should also come from owners, especially serial owners. The research on Project A is not necessarily going to help Project A. It's going to help project B, C and D.

Finally, Renée noted that just the process of completing the research has already benefited the teams simply through reflecting and framing larger questions. She observed, "the team members get so involved with the day-to-day operations and decisions—sometimes hearing a question can make them think about things differently." The teams also spent time comparing their answers to other teams and considering the differences. Once the reports are complete, they will be shared with all the teams and used for process improvement on future projects.

Future Proof

The benefits of the research will not stop within the walls of UW. Both Renée and Mike plan to publish the case study and to share the research in multiple venues. Mike will likely introduce the work at the Society of College and University Planners, the Association of University Architects, and other professional organizations. Renée will share this research in academic publications and with her network of professional organizations such as the Lean Construction Institute, which funded part of the study. Most intriguingly, both Renée and Mike will continue to build on this first step in the research related to IPD in higher education to ensure teams completing higher education projects using IPD are as effective and as efficient as possible.

Renée, however, sees an even bigger opportunity,

> We at the university have aspirations of societal change and about the people working in our buildings that are next generation—future oriented. There is potential to do things like instrumentalize our buildings and set targets not just for the building performance, but potentially for the creativity and innovation of the people living and working in these buildings. When we talk about the impact that the built environment can have on society or groups of people or individuals, we have a long way to go in our understanding.
>
> So, for me, that would be the much bigger question that we could start to ask, are there ways we can design our buildings—are there ways that we can track the way the buildings are used so that we can start to have not just successful project teams building the buildings, but success in the ways that those classrooms are working, the labs are working, and the mission of the university is achieved?

If the partnership of Michael McCormick and Renée Cheng begins to move us along the path to answering these questions, higher education and the built environment will benefit greatly from this extraordinary friendship.

Notes

1 Charles Pankow Foundation, Center for Innovation in the Design and Construction Industry, Integrated Project Delivery Alliance, *Integrated Project Delivery: An Action Guide for Leaders*, Accessed print on demand May 26, 2020.
2 Renée Cheng, "Integration at Its Finest: Success in High-Performance Building Design and Project Delivery in the Federal Sector," School of Architecture, University of Minnesota, Office of High Performance Green Buildings, US General Services Administration, April 2015.
3 Renée Cheng, "Teams Matter: Lessons from ARRA, GSA Region 5 and the American Recovery and Reinvestment Act," School of Architecture, University of Minnesota, GSA Region 5, 4240 Architecture, May 2015.

Twenty Freight Trains of Disruption   **8**

The sunlight was low in the sky as the Danish football players of Superliga and Randers FC walked into Ceres Park stadium. Long shadows of giant inflatable air dancers moved across the emerald field. Music blasted from the stadium speakers and cheering crowds could be heard above the bass notes. This day might have been any ordinary day in professional football, but this was May 28, 2020, and the crowd noises that greeted the players were piped into the stadium from recordings. COVID-19 had stopped competitive football entirely for a while. Now that play had started again, there was no chance of putting people in the stands, so the noise was welcome.[1]

As the players looked around, the seats of the stadium were mostly empty, although some had been filled with virtual spectators, paper dolls, supporting one team or the other. Surprisingly, there were some live, friendly fan faces in the stadium that day. Giant banks of monitors installed along the sidelines hosted a virtual crowd of faces, Zoomed into the stadium in one giant meeting focused on the game.[2]

It's not clear how exciting the Zoom experience was for the spectators, and there were a few technical challenges. However, European football was well on its way to figuring out how to enhance the fan experience, virtually, during the COVID-19 disruption.

Looking Out and Looking Ahead

There are advantages to living in Tucson, Arizona. Beyond the warm winters, the incredible desert, and the culinary wonders of this small but dynamic city, we are blessed with crystal clear skies most nights of the year.

So, last year, for Christmas, I gave my husband a telescope to take advantage of these skies. With this telescope, we can examine the heavens, see millions of years into the past, and track objects approaching the earth. The nice thing about these Arizona skies is, if you know where to look, you can see danger approaching.

The same is true about the design and construction industry. Though many complain that the path ahead is obscured by a foggy future, it is not hard to see what is coming at us, if you know where to look. Of course, it is hard to predict meteors like COVID-19 that seem to come out of nowhere, but there are many professionals in our industries that see clearly the need for change and are unafraid of that change. The following list outlines **twenty disruptions** industry professionals have already identified as impactful to our future and are currently exploring, testing, or developing.

1) Systems Design

Some architects and contractors have become aware contemporary problems cannot be solved in isolation. The *Economic Times* defines system design as

> the process of defining the elements of a system such as the architecture, modules, and components, the different interfaces of those components and the data that goes through that system. It is meant to satisfy specific needs and requirements of a business or organization through the engineering of a coherent and well-running system.[3]

The built environment is one component of a complex system of policies, infrastructure, business processes, and delivery. Architects like Michael Maltzan understand that some problems must be attacked systemically. For Maltzan, a rewrite of building codes in Los Angeles was needed to address the challenges of prefabrication, which, in turn, addressed the delivery of low-cost housing. Solving these problems as a system was key to meeting the goals of providing affordable housing to stakeholders in need by removing barriers to working with modular building processes and incentivizing developers to invest in those processes.

BIM has made it possible to collect and analyze data and processes parallel to the design process, optimizing the system of project delivery and the integration of decision-making. Some may say this is simply expanded problem-solving, but systems thinking adds data, and process definition to the solutions. Future design and construction professionals must think this way and develop tools to assist with a systems design approach to problem-solving in the built environment. Everything is connected to everything, after all.

2) Algorithmic Design

Some estimates state 80 percent of building elements are formulaic. According to Dictionary.com, the word "algorithm" means "a set of rules for solving a problem in a finite number of steps."[4] Architects have known for decades that there is a great deal of information being redrawn or slightly tweaked in construction documents, and, frankly, the value of the professional is diluted through this redrawing process. Algorithmic design can provide both challenges and opportunities for the architecture profession. Many architects still want to draw every project as if it is a new, original creation, yet opportunities abound in reducing redrawing, and in rapid problem-solving,

HKS uses this "set of rules" approach to design sports facilities, building skins, and to maximize shade while minimizing material,[5] an approach that will be adopted by more and more architects every year. Some architects may be fearful of being left with only 20 percent of the drawing effort. However, if professionals are not redrawing, the responsibility to test, stretch, and enhance algorithms will be critical. Architects will also have more time to focus on those areas where they can add the most value to a design, making sure it meets the user needs and public needs.

3) Machine Learning (and Artificial Intelligence)

A decade and a half ago, my colleagues at Rice University, the founders of the Academic Center for Sustainability and Energy Management, developed an algorithm that could predict the next hour's energy (electricity, gas, water, and steam) use of any building on the campus. The algorithm, based on historical usage, was weather normalized and incredibly accurate. The benefits of the algorithm included the ability to understand the benefits of recommissioning within a single day. Soon, I imagine, the software will continue to be developed so that it continuously learns not only from the weather but from occupancy patterns, equipment usage, and maintenance needs. This example is a building industry application of machine learning, a program that continuously improves formulas or algorithms using data streams. Our industry will soon leverage machine learning and, ultimately, artificial intelligence, in energy management, responsive design, and project delivery.

According to Scott Hartley in *The Fuzzy and the Techie*, "machine learning is the training of machines to perform tasks autonomously by making enormous quantities of data available to them and then programming them with a set of rules for analyzing it all on their own."[6] Machine learning has the potential to be a powerful force, but Hartley believes humans will "harness the power of big data and machine learning to assist them in tackling a host of outstanding problems that are crying out for better analytics but require both machines and people."[7] As I look at the need to solve the problems of creating more adaptable, resilient buildings that consume fewer resources, I believe we will need the power of machine learning guided by human values.

4) Augmented Reality

The patient's vital signs are showing signs of stress. Everything about the hospital environment increases that stress, including the lights, colors, equipment, smells, and sounds. The doctor, completing her examination, slips a transparent hood over the patient's face. Medical equipment melts away behind new walls and artwork. The wall colors transform, and lights soften. A window appears on a far wall revealing blue skies and a few wispy clouds. Sounds are masked, and odors freshened. Eventually, the patient's pulse slows, and stress levels are reduced. The room has not physically changed, but the patient's perception has.

This scenario might be an example of the practical use of augmented reality in the future. Augmented reality, a futuristic idea, has been made accessible to us all by an app, Pokémon, a game where we can all see little monsters in our backyard. Augmented reality, by definition, leverages existing environments, transforming them with the imposition of 3D imaging, usually through the use of a headset or other technology device. However, the use of lighting to remember the World Trade Center buildings in the Tribute in Light on March 11, 2002, was also a form of augmented reality, imposing a 3D image on an existing environment. Holograms can also be used to augment reality. I use these last examples because it is useful to expand our understanding of the approach and role that augmented reality has had and may have in our future.

Whereas virtual reality has provided us the opportunity to become completely immersed in a different world, even our design projects, before they are real, augmented reality provides the opportunity to correct imperfections, to enhance, and to collaborate in a way that virtual reality may not. In design, augmented reality will be most useful as construction commences and as clients begin to see their spaces take shape. Design professionals will be able to place elements into half-built spaces for decision-making. In heritage conservation and adaptive reuse projects, augmented reality will assist in the design, decision-making, and approval processes. During construction, augmented reality can support collaborative team problem-solving when conflicts occur.

Augmented reality will appear in other areas of our lives, including exhibits and other theme-park-like experience-making. Live performances will increasingly use technology-infused techniques to create augmented reality. Within buildings, we will accommodate art that will appear when wanted to fill a space, then disappear for an event. In a post-pandemic world, augmented reality may play an even more important role as we create working environments that enhance collaboration and personal connection even though we are connected only virtually.

This discussion brings us back to the hospital room scenario from the future. Ester Sternberg and her team at the University of Arizona Institute of Place, Well Being and Performance[8] are using wearables and cutting-edge research processes to monitor environmentally generated stress on building occupants. This exploration of person-centered health will be one of the most essential sources of knowledge to understand how our buildings impact our health. Some needs of patients will be addressed through great design or responsive design. However, augmented reality may someday also be a viable tool for addressing patient needs. It is even more interesting that we can test design solutions and investigate optimal solutions through augmented reality. We have only just begun to explore its potential.

5) Hyper-Prefabrication

Whether we want to admit it or not, in December 2011, the world changed when the Broad Group of China constructed a thirty-story hotel of prefabricated components in only fifteen days in Dongting Lake, China. The company followed that feat by building a fifty-seven-story building in nineteen days. The steel sections apparently "fit together like Lego bricks."[9] The company, founded initially to make boilers and chillers, has been thwarted by Chinese officials in efforts to build the tallest skyscraper in the world, but some sources believe it is only a matter of time until the CEO, Zhang Yue, achieves his dream. Hyper-prefabrication has already arrived as a solution to rapidly expanding cities and housing demands. The first questions are 1) who will finally determine the systems and infrastructure for well-designed major project prefabrication and 2) who will resolve the policy and building code barriers that inhibit this potentially more affordable and sustainable approach to construction?

6) Integrated Supply Chain

There have been false starts in construction industry efforts to manage the supply chain. One of the most notable and valiant efforts was the ill-fated Turner Logistics. Turner attempted to create an efficient major equipment purchasing process for multiple projects. However, the technology and available data management knowledge were not nearly advanced enough to cope with the barriers of our complex and fragmented industry.

Katerra, now an established startup delivering primarily housing and hotels, looks to be up to the challenge. This organization has no small goals as it vertically integrates the entire supply chain to achieve unprecedented results.

According to John Smith, an HBS alum,

> Katerra is fundamentally rethinking construction and is working to become an "end-to-end," vertically integrated builder. It claims that "when the entire building process is owned by a single team from end to end— bringing design, manufacturing, material sourcing, and construction together intro one streamlined system—it is possible to build high quality, beautiful buildings, faster and at a lower cost."[10]

Katerra is determined to bang out some of the efficiency in the design and construction industries. To accomplish this mission, "Katerra is building products that will enable the company to connect with workers in the field, better manage inventory, aggregate demand across multiple projects, and link its production facilities to enterprise resource planning systems."[11] Katerra not only has redesigned the prefabricated cross-laminated timber panels and integrated all building systems with just-in-time tracking of every delivery, but the company even tracks labor on the site by placing GIS stickers on hard hats. Squeezing some of the $1.6 trillion of industry inefficiency through supply chain management will be the goal of other industry players in the near future.

7) Robotics

Two years ago, I was writing a blog on disruption when I happened upon an article that startled me. It was titled, "Countdown to Human-Free

Construction in Less Than 10 Years."[12] Yes, there was a bit of hyperbole in the title, but the concept of at least one fully automated construction site being less than a decade away suddenly doesn't seem that outlandish.

According to the author, Nick Hertzman, LIDAR technologies, which make autonomous cars possible, are one of the triggers that are speeding up the adoption of AI, Pre-fab, 3D printing and, yes, autonomous machinery or robotics. Construction robots are even showing up in the form of personal assistants for workers, carrying tools, equipment, technology, and materials.

The construction industry is cyclically understaffed. The highs and lows of economic turns make the industry an unreliable industry to hold on to during the low times. In August 2019, 80 percent of construction companies in the United States were struggling to hire enough workers.[13] Hundreds of thousands of jobs went unfilled. It is no wonder contractors and subcontractors look for alternative ways to complete work and to support current workers to work more efficiently.

In spite of the challenges of earlier robotic efforts including weight, scope of work, programming limitations, and maintenance in dirty environments, robots are increasingly taking on more sophisticated jobs on project sites. Of course, repetitive activities make sense for robots such as the SAM 100 brick-laying robot, which is six times faster than a traditional mason.[14]

In November 2019, *Fortune Magazine* featured robots already at work on construction sites. Along with SAM, the article featured Hadrian by FBR building walls for entire homes in one day. Toggle fabricates and assembles rebar cages. Esko Bionics makes robotic vests for workers, supporting their arms for heavy work.[15] Self-guided robots have already arrived on construction sites, and soon, possibly in less than a decade, robots will be working, delivering, installing, inspecting, and learning on their own.

8) Mini-Robotic Construction

The tiny robot is coming on strong in research and in construction applications. I mean strong in a very real way. Mini-robots can pull many times their weight, and, when working in tandem, sometimes pull as much as 18,000 times their weight. The future application for using mini-robots to work efficiently in tight spaces, inserting sealants, applying adhesives, pulling wire, and leveling materials, is enormous. These tiny bots may even be conducting quality inspections. Mini-robots can do an excellent job moving materials around construction sites in tight spaces safely. Another advance in mini-robotics is a wall-crawling mini-bot capable of tying, looping, and handing off carbon fiber to weave intricate structures as devised by University of Stuttgart graduate Maria Yablonina.[16]

9) Swarm Intelligence

Wikipedia defines Swarm Intelligence (SI) in the following way:

> the collective behavior of decentralized, self-organized systems, natural or artificial [...] SI systems consist typically of a population of simple agents [...] interacting locally with one another and with their environment. The inspiration often comes from nature, especially biological systems. The agents follow very simple rules, and although there is no centralized control structure dictating how individual agents should behave, local, and to a certain degree random, interactions between such agents lead to the emergence of "intelligent" global behavior, unknown to the individual agents.[17]

According to Unanimous A.I., groups like bees or birds amplify their intelligence by forming swarms or flocks. People and systems can use swarming to reach decisions and generate forecasts using the collective intelligence of multiple "brains" and algorithms to sort information.[18] Researchers working on swarm intelligence are translating the simple command negotiation existing in typical swarm relationships into cooperative behavior in robots.[19] These swarm robots have multiple potential applications on construction sites. According to the Institute for Advanced Architecture of Catalonia, mini-robots may work in swarm teams to fabricate the skyscrapers of the future.[20]

10) Dynamic Decision Modeling

As noted in Chapter 4, Design and Planning decisions are becoming more and more complex. Decision-makers often do not have the necessary information to fully understand the consequences and/or ripple effects of significant decisions. Leaders increasingly understand that everything is connected to everything. Dynamic decision models connected to resource data can help organizations plan large-scale infrastructure and facilities projects more effectively, can create scenario modeling for accurate comparisons, and can be maintained/revised to be reused over time.

The Rice University Integrated Campus Plan (ICP) model, developed with KieranTimberlake is an example of dynamic modeling for scenario planning. This dynamic data-based decision model helped Rice University understand how construction impacts stormwater capacity, green roofs, housing capacity, parking, total cost, energy consumption, and many other university resources. Dynamic decision models work best when they are infused with the underlying values and "DNA" of the organization. They are not easy to build yet are worth the investment. These models will increasingly be used by planners and owners to engage in critical scenario development to plan for a more adaptive and resilient future.

11) The Trillion-Sensor Future

It was Mark Gardner of Rice University's Administrative Center for Sustainability and Energy Management (ACSEM) that first mentioned to me the possibility of the trillion-sensor future for the built environment. Mark and I worked together for years at Rice, and I was always impressed by his ability to leverage data from wherever it came. It started with building controls, energy meters, and plant optimization systems. Mark and his colleagues at ACSEM built algorithmically driven systems maximizing data knowledge by predicting the next hour's energy use of a building (weather normalized) for troubleshooting, recommissioning, or reshaping energy load consumption to change purchasing structures.

Additional sensors are now placed in buildings to support responsive design systems, increased operational needs, sophisticated preventative maintenance systems, expanded security systems, and even to meet building research needs. It's not just about energy management anymore; it's about the data.

There are many questions unanswered at this point. How will all these systems be integrated? Who will design these systems? Who will coordinate and install them? Who will own the data? Who will know how to control the data? Who will be able to imagine how those data are best leveraged?

At the 2018 Design Futures Council Leadership Conference in La Jolla, California, on Design Technology & Application, the idea of a CyberQuake was introduced by Angela Watson of Shepley Bulfinch. Angela asked what the role of the design and construction professional would be in a world where a building might be attacked or hacked, causing a "cyber quake." A "cyber quake" is defined not only as a building that stops working but also as a building that might start attacking its occupants when its sensors send the wrong messages about temperatures, sun angles, time, CO_2 levels, and so on.

The trillion-sensor future provides opportunities to adapt and treat aging occupants, respond to a pandemic, and shape environments for personal comfort. The trillion-sensor future also provides opportunities for data collection and extraordinary research. However, the trillion-sensor future raises privacy and security concerns for which we must anticipate and prepare.

It's the very latest collaborative fully immersive virtual AR big data analysis enterprise system.

12) Property Technology and Design Technology

Just a few years ago, the PropTech (property technology) industry was just a $5 billion industry. Today it exceeds $18 billion.[21] These technologies are developed because real estate developers and owners need better, more transparent information to buy, own, manage, build, and maintain their properties. As noted in Chapter 4, design and construction industry professionals should expect to comply with the rules of this new technology, be ready to answer the questions asked by this technology, and change processes to integrate with this technology. Take note, one of the reasons for the growth of PropTech is that venture capitalists are interested in the data gathered through these technologies.

Alternatively, Design Technology is evolving not just to support collaboration and to ease the process of design but now to deliver knowledge to designers just in time to make great decisions. KieranTimberlake's Tally® Life Cycle Assessment App[22] is one of those tools that can support designers with critical knowledge about embodied carbon, energy reduction, and other environmental impacts at the critical moment of decision-making when it can have the greatest beneficial impact on a project. In time, PropTech and DesignTech will be deeply integrated.

13) The Competition for Talent

Before March 2020, everywhere I turned in the design and construction industries, I heard about the need for talent. The growth of the market before the pandemic was one aspect of this challenge. Another challenge was the industry's cyclical character, impacting the number of people hired from year to year, from decade to decade.

However, over the last decade, we have faced a triple challenge. The generation of millennials following the boomers has been too small to take over the middle management responsibilities. The talent loss in the design and construction industries experienced after 2009 and the Great Recession was devastating. Finally, the reluctance of the Baby Boomer leaders to let go of the reins of leadership has created a gap in experience in the core of our industries. Repeatedly, over the last two years, I am hearing firms forced to give project leadership to completely inexperienced Gen X professionals. This dynamic has blessings and curses that are manifest for the industry. These young professionals will have a chance to be super professionals, but they will make many mistakes.

COVID-19 will have at least two effects. Many Boomers will leave the industry in accelerated retirements, and more young talent will be lost in the pandemic downturn. The inevitable result will be an even dimmer talent picture when the economy picks up again.

14) Loss of Licensure

The danger of architects becoming extinct is most tangible in the recent push against licensure by states and other professions. This concern has been bubbling up in multiple states. It has manifested in bills that have challenged licensure for interior designers, landscape architects, architects, and even some plumbing and mechanical trades. This movement is caused in part by YouTube making us all feel that we can design our own home or plumb our disposal—if only we can find the right video. It is more likely that roles on projects are getting deeply blended. It is not clear that the definition offered by the National Council of Architectural Registration Boards (NCARB) defining the skills of an architect are the skills needed in today's industry or in the industry of the future. The same goes for engineers, landscape architects, plumbers, and many other professionals. Perhaps we should consider that losing the current rigid constraints of licensure might help the industry and its professionals.

15) SuperUsers

I am adding Randy Deutsch's book, *Superusers: Design Technology Specialists and the Future of Practice*,[23] to my list of industry disrupters. This book, written about the importance of design technologists, with specific characteristics that make them "Superusers" to the architectural firm is a book that defines the key twenty-first-century professional for much of the design and construction industries.

Ian Keough recognizes in the book's Foreword,

> The Superuser is [...] a heroic character as she fights the often backwards and inept processes by which buildings are designed and constructed, doing so from within [...], and considers a different future as he wonders, "What will our profession look like when the best among us, those who are asking the fundamental questions about the place of technology in architecture, are the leaders?"[24]

This is the crux of the matter. As Randy describes this new architectural professional blessed with the ten C's, including the ability to Contextualize, Communicate, Concentrate, and Connect, a desire to Collaborate and Continually improve, blessed with Curiosity and Capacity, and skilled in Computational thinking and Coding, it is hard to imagine any business where such a person would not be valuable and, potentially, a leader. In architecture, where the inspirational designer has traditionally been tapped first as the partner or head of the firm, a strategy that is failing in influence and impact in 80 percent or more of the world's constructed environment, we must consider alternatives. It is clear the leaders of our industries will be molded from different clay in the next decade and they must have a grounding in technology to be effective. Deutsch has dared to define that professional and I like what I see:

Superusers are design professionals with the wherewithal to recognize a tool, curiosity to inquire into a tool, confidence to mess with a tool, capacity to combine tools, and the interpersonal intelligence to connect with others to achieve actionable results.[25]

As I wrote in a review of Randy's book, I wish I had a hundred copies to give to educators and leaders in every corner of our industry. This is how we should be thinking about talent. This is how we can help young professionals create a successful future.

16) Super-Teams

Amy Edmondson from Harvard has been focused on the power of teams in her research for many years. In her book with Susan Salter Reynolds, *Building the Future: Big Teaming for Audacious Innovation,* Edmundson examines the urgent need for super-teams that are high performing and multidisciplinary to address audacious problems of the future.

The authors note,

Future-building is hard. When success requires introducing what Machiavelli, in the sixteenth century, called "a new order of things," success is likely to be elusive. This is because bringing together diverse elements (technologies, plans, people, or organizations) to create a functioning whole presents countless ways for integration to break down.[26]

To accomplish these challenges, the authors argue we need, "a new kind of collaboration that spans more (and more diverse) groups than ever before."[27] If you think collaboration is hard now, try collaborating on a Super-Team.

17) Market Forces/Data Value

As noted earlier in both the Trillion Sensor Future and the PropTech sections, the value of data has come under scrutiny by venture capitalists. Built environment data is valuable because it will drive markets in health, products, materials, responsive design, and resources. We cannot yet imagine how access to building data, generated by building access, elevator trips, videos, security devices, system sensors, lighting controls, and human response, will be useful. However, venture capitalists understand that he/she who owns that data will hold a significant asset. Smart designers will find ways to retain ownership of data for their clients or themselves.

18) Market Forces/Cost of Construction

The cost of construction has continued to escalate at a rapid pace. Prices are so high, building is becoming unaffordable and unachievable to small business owners, young homeowners, and nonprofit organizations. Buying and maintaining the built environment is becoming untenable, especially when one considers the resources consumed. Other disrupters are entering the markets just because of the market strain caused by the cyclical economic weeding out of small suppliers during recessions.

The pandemic has shown many of us we can work from home, potentially disrupting the real estate market. An excess of space in urban areas combined with the high cost of new construction could create a severe reexamination of the design and construction industries.

19) The Startups

Millennials, Generation X, and Generation Z are unafraid of the world built by Baby Boomers. They are happy to start up their versions of the design and construction industries and genuinely interested in changing the world. This entrepreneurial tendency does not resemble the small architectural firm of the last century. These startups are intended to have an impact and to make a difference. Already referenced, Katerra, WeWork, and Unispace are all companies that are or were reinventing and integrating the industry in different ways. They are joined by PropTech, Robotics, PreFab, and equipment-sharing entrepreneurs. Whether these companies succeed or fail, they teach and inspire new and different startups that will succeed in disrupting our industries.

20) Crowd Manufacturing

There was a moment during the pandemic, May 30, 2020, when many of us stopped, held our breath for a few moments, and then smiled. It was the first genuine smile in quite a while, but seeing the astronauts Bob Behnken and Doug Hurley safely into space on the first American launch with humans aboard since 2011 was satisfying. This launch was a very different experience, however. This launch was not by NASA. It was by Elon Musk's private company, SpaceX.

In January 2020, just before the pandemic hit the United States, an intriguing book, *The Future Is Faster Than You Think*,[28] was published. By Peter H. Diamandis and Steven Kotler, the book introduces us to the many ways the world is changing much more quickly than we may believe possible, mainly through the efforts of independent entrepreneurs. In the 1960s, the space mission, driven by government investment, drove extraordinary engineering and technology advancement. Today, that advancement is coming in different forms and on different paths.

Innovators like Elon Musk inspire us to believe that we can all act and make a difference. The pandemic brought into focus another kind of entrepreneurial innovation with immense power to change the design and construction industries, crowd manufacturing. When personal protective equipment (PPE), face shields and masks, were needed desperately by hospitals, large manufacturing companies and supply chains could not pivot quickly. These items were manufactured in university maker spaces, plumbing shops, designer sewing rooms, and private homes. Small teams of inventors popped up everywhere around the world to develop better and cheaper

Figure 8.1 MODI Homes, a Texas start-up, incorporates prefabrication and crowd manufacturing into its steel frame, modifiable, moveable modular home, modihomes.org
Photo credit: Jacob Miles

ventilators, masks that didn't hurt ears and faces, and intubation boxes that protected medical workers from contamination while allowing then to see clearly. These inventors shared patterns, models, and technical details. They brainstormed solutions and helped adapt equipment to new uses.

The power of the small 3D printer, the computational modeling tools, and the accessible, sophisticated maker tools available to many designers and builders have changed the game. We can now imagine, make, ship, and sell. The Asian manufacturer and the complicated supply chain can be pushed out of the picture for some products. Look for more designers and builders edging into crowd manufacturing opportunities now we know we can make it work. It doesn't take the government to build a spaceship. It doesn't take GM or Toyota to deliver PPE or even to build a car anymore. Creating a new product and delivering it to the market may only take a little crowd manufacturing.

Summary

No traditional contracts, delivery methods, or education can prepare industry professionals for these disruptions. However, you can prepare for the future first by examining your organization for strengths, values, and goals. Where those strengths, values, and goals align with a potential disruption, you have an opportunity. Do not choose all twenty. Do not be afraid of all twenty. Find one or two, maybe three, disruptions you believe may align with your organization and then commit to preparing for those opportunities. Do not hesitate and do not resist change. As Barbara Ward, a famous British economist, once wrote, "It is a fact of history that those who seek to withdraw from its great experiments usually end up being overwhelmed by them."[29]

High-performing teams can help you prepare and cope with disruptions, both expected and unexpected. A Culture of Predictable Outcomes can create radical solutions to the design and construction industries' most challenging problems.

Notes

1 Mike Koeshartanto, "Soccer Team Uses Zoom to Create Virtual Crowd," *Gilt/ Edge Soccer Marketing*, May 19, 2020 (updated May, 29, 2020), www.giltedgesoccer. com/soccer-team-uses-zoom-to-create-virtual-crowd/.

2 David Fucillo, "Danish Soccer Club AGF Aarhus Has Zoom Crowd 'in Attendance' at League Restart," *DraftKingsNation*, May 28, 2020, https://dknation.draftkings. com/2020/5/28/21274056/agf-aarhus-vs-randers-fc-danish-soccer-leaghas-zoom-crowd-in-attendance-at-league-restart

3 "Definition of 'Systems Design,'" *Economic Times,* https://economictimes. indiatimes.com/definition/systems-design.

4 "Algorithm," Dictionary.com, October 23, 1998, www.dictionary.com/browse/ algorithm?s=t

5 "Building With Data: How HKS' LINE Innovation Team Drives Change Using Algorithms, Creativity, Nature," Dallas Innovates, October 30, 2018, https:// dallasinnovates.com/building-data-algorithms-creativity-nature-drive-change-hks-line-innovation-team/.

6 Scott Hartley, *The Fuzzy and the Techie: Why the Liberal Arts Will Rule the Digital World* (Boston: Mariner Books, 2017, Kindle ed..), 37.

7 Ibid., 41–42

8 The University of Arizona Institute on Place, Wellbeing & Performance, https://ipw.arizona.edu

9 Jamil Anderlini, "Zhang Yue, Broad Group—China's Flat-Pack Skyscraper Tycoon," *Financial Times*, November 20, 2016, www.ft.com/content/28e22386-a11c-11e6-891e-abe238dee8e2.

10 John Smith, "Building a Better Way to Build at Katerra," Technology and Operations Management, HBS MBA Student Perspectives, November 15, 2017, https://digital.hbs.edu/platform-rctom/submission/building-a-better-way-to-build-at-katerra/.

11 Ibid.

12 Nick Hertzman, "Countdown to Human-Free Construction in Less Than 10 Years," ForConstructionPros.com, January 10, 2018, www. forconstructionpros.com/profit-matters/article/20987766/countdown-to-humanfree-construction-in-less-than-10-years.

13 Jennifer Alsever, "How Robots Are Changing the Construction Industry," *Fortune*, November 23, 2019, https://fortune.com/2019/11/23/construction-industry-robots-ai/

14 "Construction Robotics Bricklaying Robot SAM100-2016," YouTube, April 8, 2016, www.youtube.com/watch?v=mP2AtqcitbQ&feature=emb_title

15 Jennifer Alsever, "How Robots Are Changing the Construction Industry," *Fortune*, November 23, 2019, https://fortune.com/2019/11/23/construction-industry-robots-ai/

16 Rima Sabina Aouf, "Wall-Climbing Mini Robots Build "Entirely New Structures" from Carbon Fibre," *de zeen,* August 2, 2016, www.dezeen.com/2016/08/02/wall-climbing-mini-robots-construction-carbon-fibre-university-stuttgart-achim-menges/.

17 "Swarm Intelligence," *Wikipedia,* https://en.wikipedia.org/wiki/Swarm_intelligence.

18 "How Does Swarm Work?" Unanimous AI, https://unanimous.ai/what-is-si/.

19 "Controlling Robotic Swarms," YouTube, December 2, 2014, www.youtube. com/watch?v=stzQNjtDg0g.

20 Adele Peters, "This Band of Small Robots Could Build Entire Skyscrapers without Human Help," *Fast Company*, June 19, 2014, www.fastcompany.com/3032003/this-band-of-small-robots-could-build-entire-skyscrapers-without-human-help.

21 Gary Barker, "Is PropTech about to Have Its 'Man On The Moon' Moment?" *Forbes*, December 16, 2019. www.forbes.com/sites/garybarker/2019/12/16/is-proptech-about-to-have-its-man-on-the-moon-moment/#69d6ec46ec02.

22 "Tally® Life Cycle Assessment App," KieranTimberlake https://kierantimberlake.com/page/tally

23 Randy Deutsch, *Superusers: Design Technology Specialists and the Future of Practice* (Abingdon, Oxon, UK: Routledge, 2019).

24 Ibid., xi–xii.

25 Ibid., xvii.

26 Amy C. Edmondson and Susan Salter Reynolds, *Building the Future: Big Teaming for Audacious Innovation* (Oakland, CA: BK/Berrett-Koehler, 2016), 6.

27 Ibid., 8.

28 Peter H. Diamandis and Steven Kotler, *The Future Is Faster Than You Think* (New York: Simon & Schuster, 2020).

29 Joanne Kelly, *The Gigantic Book of Famous Quotations: Over 12,000 Famous Quotations to Inspire, Motivate, Comfort and Amuse You!* (UK, Kindle ed., 2019), loc. 14923.

Chapter Supplement—8
Scott's Choice

Optimism Is True Moral Courage

It would have been simpler to give up and to go along. Scott knew he was marching against the wind, but he absolutely knew that the information before him was not correct, that it did not tell the entire story. No one would ever challenge his decision if Scott just accepted the material as presented and just went along. But would he be able to live with it? On the other hand, if Scott fought the conventional wisdom, it would be, to cite another aphorism, a classic uphill battle. The cost estimate before him proved it, and the construction schedule was unforgiving.

"It's my keister! I'm going for it," Scott thought to himself. Out loud, he said, "These numbers are wrong. We aren't done until I'm satisfied that you understand this process—that you understand what construction with prefabricated walls will mean to your schedule and costs—that you understand what DIRTT can do!"

Perhaps, at that moment, Scott Shackleton, assistant dean for capital projects and facilities for UC Berkeley's College of Engineering, had embodied the leadership style of his famous ancestor Ernest Shackleton, the Antarctic explorer. It was Shackleton who often said, "Optimism is true moral courage,"[1] an idea Scott embraced. However, there was a great deal, at the moment, that could encourage pessimism.

The goal of the 27,000 square foot project was to house the artificial intelligence program for the college. Scott had DIRTT, a prefabricated wall system in mind from the beginning of design, but had encountered nothing but barriers implementing this solution.

A Different Kind of Leadership

Scott was not your typical owner's representative when it came to the design and construction industries. He came up through the university's central facility management office. However, the dean of the engineering college wanted someone to steward the college's interests and money, so he asked Scott to help. The college of 6,000 students and almost 2 million gross square feet was nearly the size of a small private university, anticipating a significant expansion, so the challenges were unusual and intriguing.

However, from his first project at the college, Scott was discontented with the traditional delivery processes and with the adversarial relationships between industry partners. As a result, Scott endeavored to use his Naval Reserve background in leadership to shape different experiences and procedures on the projects he led during his years at the engineering college, emphasizing collaborative engagements and quick, effective, problem-solving-oriented decision-making.

Scott is also profoundly engaged in team dynamics.

> It is so important to create relationships within the team members. You know, I can honestly say that every project I worked on since then, I'm still good friends with everybody on those teams—general contractors, architects—and I use different architects on almost every project. But we've created strong relationships and trust. And it's such a huge component for me. I feel that's a huge component of this. It's been part of my mantra for all my projects. Scott observed teaming includes tough choices,
>
> When I've had team members that don't fit that group, don't want to cooperate, don't want to be part of the team, I get rid of them and I mean, it's difficult, but I've done it a number of times. It could be a member of the general contractor's team, a member of the architect's team, I've even gotten rid of members of the UC team. If they're not all in, then they're out, and that's the way I roll, and people know that.

The DIRTT Secret

Scott also knew the DIRTT product well. He had been using it incrementally for several years on smaller projects. DIRTT had been the right choice for some challenging projects in the past where installations have been required in short time frames or in areas where traditional construction was not advisable. Scott's first experience with DIRTT was a library renovation where book stacks had to be replaced by five small conference rooms, two large conference rooms, and one training room during summer break, a period Scott refers to as "72 days of hell." In this case, the architects recognized the opportunity for a prefabricated product and recommended DIRTT. Scott responded, "What the heck is DIRTT?" The architects were happy to educate him on the product.

Scott remained skeptical and worried about acoustical challenges in this environment but was surprised and even a bit shocked by the fast and accurate installation. Seven rooms were installed in five days. These rooms were completely finished, compared to what would have been four weeks and a punch list on a traditional project. Scott marveled at the fact that only installers were required on DIRTT versus "lots of different crafts coming in, you know—studs, drywall, tape, painting, and one guy who can't make it in because he got held up on the job and lost three days waiting on somebody."

Scott also recognized the quality of the work was different when using DIRTT. "Everything's precision cut, everything fits, everything shuts, it's not awkward [...] I was just blown away, and I like I kind of drank the Kool-Aid. I said, 'Wow, this is the wave of the future.'"

As Scott developed more jobs with DIRTT, he recognized other benefits. Procurement was not through a construction contract. Instead, DIRTT was procured through furniture purchasing processes and contracts. The DIRTT ICE® design and estimating program gave Scott an accurate price, even in the earliest phases of the project, and DIRTT's prices never went up, so no contingency was required for DIRTT. DIRTT installation environments were clean, dust free, and overall, safer. After a bumpy start with plan reviewers and field inspectors, reviewing and inspecting DIRTT had become predictable and straightforward, even if the project teams using DIRTT were being held to a higher standard by inspectors than for traditional construction. And Scott was beginning to realize, his first impression, that DIRTT was an expensive solution, was not precisely correct. Once risk was reassessed and contingencies reduced, once redundancies were removed from other parts of the estimate, the DIRTT product was often comparable in price—for a better quality product with greater flexibility.

Scott indeed "drank the Kool-Aid" related to DIRTT when he visited their manufacturing facility in Canada.

> When I saw it, their whole ICE® software and how it works [...] it's like, oh my god this is crazy, this is amazing, it's like a massive 3D printer for construction. That was really where the sword went through the heart. I saw they were building it out clean. Their computer organization—it's kind of like Legos, you just put the blocks together, and they're all perfect and everything lines up.

So, when Scott saw an opportunity to use DIRTT on an entire floor of a research building for the college's AI team, he jumped at the chance. This project would replace every part of the tenant improvements (TI) with

DIRTT. Every part of the interior space, except the core, would be built out with DIRTT walls, including specialty areas like the café. Scott believed he could create a better, more flexible environment for the same price using DIRTT. Scott had taken the AI researchers to Silicon Valley to see typical DIRTT installations and had received enthusiastic agreement from the future occupants. However, Scott knew that DIRTT would never be an option if it turned out that the prefabricated product cost more than traditional construction. He could never go to his CFO and argue that it was a better value if it cost more.

Decision-Making

Scott chose an architect and estimator based on preliminary conversations that made him believe they were open to using prefabricated products. Both companies stated they were willing to use the product, at least until the time came to do the drawings and the estimate. Suddenly, both the estimator and the architect seemed resistant. Well, maybe not resistant, possibly clueless and incredibly risk averse. For example, in his first number, the estimator had built in a 20 percent contingency, a million dollars of contingency. From Scott's perspective, this was unreasonable since most of the project cost was the DIRTT products and installation, and those prices were, as Scott said, "Rock solid." Scott went on to say, "Whatever general contractor I got, they were going to pay that the DIRTT price [...] this is totally predicted. This completely takes the worry out of going through the bid process."

When Scott and his project manager, Case, analyzed the cost estimate, it was clear the estimator did not understand the DIRTT product. Significant redundancy existed in the estimate between the DIRTT product services and the contractor services, including electrical, installation, and other construction activities. Scott and Case sat down to educate the estimators, architects, and the contractor about DIRTT, walking through every aspect of the product and the installation. According to Scott,

> We had these workshops with DIRTT. We brought in all their experts. We sat down with the estimators, don't put this in. Oh no finish [...] no sheetrocking. There's no painting—there's no electrical, all you are putting in are these junction boxes under the floor. We're going to be plugging into these, and these are going to be provided by them. The electricians are just running home runs to the boxes. It was a whole new thing for them.

By the time the team was finished with the work sessions, the comparable estimates between the DIRTT construction and the traditional drywall construction were virtually even, possibly even less expensive for the DIRTT project. However, this process took time, and the project was now officially behind schedule, unless, of course, they chose to move forward with DIRTT. This time, Scott said what he was thinking out loud, "It's my keister! I'm going DIRTT."

Procurement and Other Problems

Scott received approval from the college leaders and moved forward as quickly as possible, but procurement or contracting was a challenge. Contracting took longer than it should have because of the lack of understanding of the product and process. Once the team was genuinely moving forward, Scott found himself still fighting a bit with reluctant architects. They didn't understand that drawings by DIRTT reduced the need for many architectural drawings and reduced the need for submittals. Many submittals were completed before the contractor was even on board. The three months lost on the DIRTT decision, and related contracting was made up almost immediately as the installation schedule was defined.

A few problems arose. For a while, university IT staff insisted on conduits being installed in the DIRTT panels until Scott demonstrated the ease with which the panels would be removed. Conduits were simply a wasted expense with this level of wall access. In addition, the contractor mismeasured the elevator height, so the panels, manufactured in Canada to reach the tall ceilings would not fit into the elevators and had to be hand-carried eight floors. The panels could have easily been manufactured to fit in the elevator and then assembled on the floor.

Outcomes

In the end, the contractor project manager was amazed at the lack of dust and rework required by the DIRTT product, which was being installed much faster than he could imagine. His biggest challenge was logistical. There were carts everywhere with panels and product. The contractor planned to design a staging and delivery process on the floor next time to avoid the cart traffic jams they were currently experiencing.

Again, Scott was worried about acoustics in the project, but the rooms were already testing at STC ratings comparable to drywall environments, so he was satisfied that sound transference would not be a problem. Faculty were delighted with their spaces, although one researcher regretted a decision he made that blocked his view of the Bay Bridge. Scott contacted DIRTT to see what it would take to replace a couple of panels with glass and told the faculty member that after he moves in, if he still prefers a view, for $4,000, they can change out those panels before lunch one day.

In the end, Scott spent about half of the owner's contingency, $250,000 of $500,000, on scope improvements, mostly on underfloor air handler enhancements. Very little contingency went to unforeseen conditions.

The project opened on time and under budget. It was completed several months more quickly than the Microsoft project that had started construction at the same time on a lower floor. Scott is already looking at some upgrades he would like to make in the space, including incorporating AV directly into the wall panels, something they did not have time to do since the AV decisions were made so late. It won't be difficult to incorporate these changes after completion, thanks to the flexible DIRTT product and, possibly, to Scott's keister.

Note

1 Caroline Alexander, Frank Hurley, "Epic of Survival:Shackleton," *National Geographic Magazine* (Nov. 1998), Vol. 194, 90.

Unpredictable   9

Bad Things Happen

I can hear the voices of many readers as you complete this tome. Every experienced professional in the design and construction industries is yelling at me right now, "Projects are unpredictable! Bad things happen!" I agree. If I have learned anything after forty years in design and construction, these industries are amazingly unpredictable—bad things do happen.

This year, 2020, we are facing a pandemic, among other challenging events. In 2009, we faced the Great Recession devastating the economy, crashing construction prices, and draining our young talent pool. In 2000, my teams and I weathered Tropical Storm Allison in Houston, which filled the unfinished basements and parking garages of six projects, all storing equipment and materials, with millions of gallons of putrid water. In 1986, I had a $75 million project put on hold two weeks before the end of construction documents, and then restarted only to be completely redesigned. To my everlasting regret, a man lost his life in a construction accident during a microburst on a project under my direction.

Unexpected things happen in this business. That is precisely why building a Culture of Predictable Outcomes is so important. When I coach teams across the country, regardless of the process, for owners, architects, and contractors, I encourage them to set up relationships, skills, tools, processes, and cultures that will serve them well, especially when things go wrong. While it is important to recognize that external challenges will continue to make our lives interesting, it is the unintended challenges and barriers we create for ourselves that we can attack and resolve within a Culture of Predictable Outcomes. This result is often the most surprising aspect of my coaching experiences. Teams generally do not realize how often we shoot ourselves in our own feet, and how easy it is to avoid.

The Good Stuff

It turns out, though, that as effective as sophisticated-caring leadership, high-performing collaborative teams, and master-level decision-making can be, the most potent stuff actually happens at the intersections of the three elements.

Leadership + Collaborative Teams = Effective Collaboration

When sophisticated-caring leadership combines with high-performing collaborative teams, collaboration is super-charged and even more effective. Collaboration moves to the next level when leaders have a light touch, stepping in only when teams need influence or help to make sense of a problem. As Carraher, Smith and Delisle note in *Leading Collaborative Architectural Practice*, working in this fashion, "team members learn, grow, and gain competency."[1] Leaders deeply committed to making teams work, nurture and protect the collaborative environment, continuously building team effectiveness and success.

Decision-making (Data) + Collaborative Teams (Humans) = Better-Informed Timely Decisions

These days, big business is often obsessed with Big Data. The design and construction industries will have to adopt Big Data approaches if we are going to be better decision-makers in the future. However, data do not tell us everything we need, to create decision-ready information. Data need analytics and human perspective to be valuable. Ben Reiter explained how this applied to baseball in *Astroball*. After *Moneyball*, the book that described how analytics helped the Oakland Athletics win, traditional talent scouts were discounted. But in Jeff Luhnow's organizations,

> It's the scouting information and the performance information. A systemic rejection of all potentially predictive information the scouts provided, based on their experience and observations, meant discarding potentially talented babies along with the bathwater. The scouts might be flawed—they were, after all, human—but they still had value. It was for this reason that in 2006, Sig [Mejdal] had developed the first iteration of a metric that sought to incorporate the reports of the club's scouts with his own performance-based algorithms, to integrate

quantitative and qualitative evaluations. He called it STOUT—half stats, half scouts.[2]

It turns out that humans plus teams can inform data and decision-making deeply and effectively. The Strata Institute report on the Future of Work, *Robot Ready: Human+Skills*, reinforced the need to have humans translate technology and data, "The skills needed now and for the future combine the technical with the human: programming + ethics, artificial intelligence (AI) + emotional intelligence, or logic + values or judgment."[3]

When we combine master-level decision-making discipline with high-performing collaborative teams, we leverage our ability to make better-informed and timely decisions.

Leadership + Decision-Making = Project Value

In *The Owner's Dilemma*, I spent many pages on decision-making and the owner's role in shaping effective decision-making. "When decisions are made at the most powerful moment and the most powerful (efficient!) level within the team, great value is generated and potential for the owner to gain great value is maximized."[4] Leaders can reinforce decision discipline and ensure processes are extensively and correctly used and maintained. When sophisticated-caring leaders combine with master-level decision-making discipline, value is created, sometimes enormous value.

"How many contractors does it take to change a light bulb?"
I looked at him and grinned, "Change? Contractors?"

An Innovative Future

When I started writing this book, the Astros were World Series winners, but they had not yet been caught up in the sign-stealing scandal. I read Ben Reiter's quote in 2018, stating the Astros just want to make a few decisions better than the next team. Reiter has said, "The Astros, though, never claimed to own a crystal ball, or that they would never make a mistake. They always expected to make many of them. Their goal was to make marginally more correct decisions than their competitors."[5] At the time, I thought, well, that's okay for the Astros, but that's not good enough for the design and construction industries. We can no longer be content with incremental improvements. We must create radical changes in the way we do our business. These changes will only emerge from a different culture that can overcome our fragmented industry and fully integrate our work. Only then can we innovate and create new and more rewarding ways to deliver the built environment.

A Culture of Predictable Outcomes provides a path to making you and your teams more capable and resilient. A friend and respected colleague of mine, Mike McCormick, from the University of Washington, told me once that delivering great projects is relatively straightforward. You just have to follow a particular road map. I agree with Mike. The process is not hard to figure out, but it is hard to follow because many forces are trying to push us back to bad behaviors and bad practices.

I hope creating a Culture of Predictable Outcomes helps leaders interested in collaborative projects and excellent outcomes follow a straight-forward path to success. I hope this book, outlining leadership, teaming, and decision-making, supported by values, helps teams stay on the path to meet their goals. It is hard. As Larry Keeley noted in *Ten Types of Innovation*, "The problem with trying to change the culture of an organization is that it's a bit like trying to hug a cloud—you can see and feel it, but it's hard to get a grip on it."[6] However, nothing supports change as effectively as giving new ideas a try. As Larry also states, "It turns out that when your people act and think differently over time—and when they see different and better results emerge from these behavior shifts—culture takes care of itself."[7] In the parlance of a popular shoe manufacturer, "just do it."

For us in design and construction, we can and must set the bar higher. We are so DARN inefficient and so unpredictable that we can realize massive improvements by prioritizing culture rather than process. Collaborative teams that are well led and capable of making great timely decisions will innovate and create fantastic processes we cannot now even imagine. By creating a Culture of Predictable Outcomes, you and your teams will get the winning outcomes you want and predictable outcomes our industries need.

Notes

1   Erin Carraher, Ryan E. Smith, and Peter Delisle, *Leading Collaborative Architectural Practice* (Hoboken, NJ: Wiley, 2017), 137–141.

2   Ben Reiter, *Astroball* (New York: Three Rivers Press, 2018), 28–29.

3   Strada Institute for the Future of Work, Emsi, *Robot Ready: Human+Skills, The Future of Work,* January 16, 2019, www.stradaeducation.org/network/institute-for-the-future-of-work/.

4   Barbara White Bryson and Canan Yetmen, *The Owner's Dilemma: Driving Success and Innovation in the Design and Construction Industry* (Atlanta: Ostberg Library of Design Management, Greenway Communications, 2010), 104.

5   Ben Reiter, "About That Prediction … How the Astros Went From Baseball's Cellar to the 2017 World Series," *Sports Illustrated,* October 24, 2017. https://www.si.com/mlb/2017/10/24/houston-astros-sports-illustrated-world-series-prediction

6   Larry Keeley, Ryan Pikkel, Brian Quinn, and Helen Walters, *Ten Types of Innovation* (Hoboken, NJ: Wiley, 2013), 200.

7   Ibid.

Chapter Supplement—9
Culture Shift

Note: Names of participants, firms, and institutions have been changed at the request of the participants.

Losing

Pierce just could not believe his eyes or ears. Actually, he could. Pierce had predicted this exact outcome, and it could have been avoided. He watched as a perfect presentation crumbled before his eyes, and the company's chances of winning this project died. The superintendent was reading his script, a script handed to him by the director of marketing an hour ago. The superintendent, Ben Barnes, was one of the best in the business, and was always in control and composed on the job site. But here, in this room, standing in front of strangers in an ill-fitting suit, Ben looked small and inept. Worse, he read the script as if he had no idea what he was saying or doing.

In his sales and team manager role, Pierce had been working for years to get a chance to present to this client on a major project. Their construction company, Dynocon Construction, had an excellent reputation. It was multigenerational and had been in the community for a long time. It had the experience and the capacity, but this client had a traditional stable of other contractors that they had always used. But this was a great client with a great deal of work, so getting on their list of preferred contractors would be a tremendous boost for Dynocon.

After the superintendent's speech and after the questions and answers where Ben uncomfortably answered safety questions and logistics questions in cryptic sentences, the owner's team was not impressed. This was the man that was going to build a critical building with $30 million of their money? They would be working with this man for years. No way, their faces said, could this guy get the job done. All that work had gone up in smoke in just a few minutes.

Pierce reflected on the events of that day and the days preceding the project opportunity's devastating loss. This loss was not Ben's fault. The team was responsible, Pierce decided. The team had met three times to practice the presentation. The director of marketing, Jim Turkish, decided he would be in charge even though this was Pierce's client. Jim decided that the only people needed at rehearsal were a senior leader, the project executive, the project manager, and himself. Pierce had to argue that he would be team manager on the project just to be able to stay in the rehearsals. Pierce argued to have Ben and the project estimator attend the rehearsals, but Jim insisted there were already too many people involved, and they were busy on their current jobs. Their parts were small anyway.

The rehearsals were tortuous. Jim wanted to second-guess every single image and speech. By the end of each session, every participant was on edge and exhausted. During the last session, Jim insisted on creating a list of potential questions the client might ask and spending hours dissecting answers for each question.

Ben was not part of any of these conversations. Yet, Ben and the estimator, Doris, were told to show up for the presentation and give a short speech each on job site safety and project estimating respectively just 30 minutes before the team was invited into the room to set up. At that point, Ben and Doris knew almost nothing about the project or the people to whom they were presenting. Doris seemed to take this in stride, but Ben shifted from foot to foot reading and rereading his script.

Ben wasn't the only one that seemed uncomfortable, however. The presenters would be a senior vice president, a vice president, the project executive, the project manager, the project estimator, the team manager, and the project superintendent. Jim was not part of the presenting team and was, therefore, not invited into the room. Despite the rehearsals, the project executive and the project manager seemed ill at ease. As the team set up the easels, they seemed to be riffling through their notes in some confusion. Pierce hurried over to help but was caught up by the senior vice president. "I know you are supposed to open, Pierce, but I am the senior person on the room, so I think I should." Pierce was stunned for a moment but not for long. "Of course," he said, "I'll pick up after you to connect the dots." Pierce noticed that the client team was beginning to sit down, and he hurried to the PE and PM still looking uneasy in the corner.

"What's up?" Pierce asked.

"I think the project manager should talk about high-rise construction, not the project executive," stated the PE leaving the unprepared PM looking miserable.

"Too late to change," Pierce hissed as the senior VP opened with a cheery "Good Morning!" Pierce peeked at his watch. It was 1:30 p.m.

Despite the semi-chaos, the presentation had gone pretty well. The senior VP was mostly a non-factor. The client was simply not interested. Pierce then reassured the client team, with which he had built a good relationship. If Dynocon Construction were chosen, Pierce would continue to be involved as team manager. He described the role as general problem solver and client liaison. The rest of the team then presented their rehearsed portions of the presentation. The PE seemed uncomfortable with the high-rise construction section, and he wouldn't have managed questions well if they had come to him, but the rest was pretty good. The project manager, relieved to be presenting items he had prepared, was professional, although wooden and distant. Doris, the project estimator, threw away her script and gave the same canned speech she always gives at every presentation. It always worked too. She knew her business. Then came Ben and his downfall. Pierce's stomach still clenched as he remembered how Ben's hand trembled and how his large frame seemed to shrink into the room. He had seen Ben manage six teams in a crisis without missing a beat, but in that room, he was completely unarmed.

There were many things that Pierce imagined could have been better about that day and that presentation, but he knew that one element must be addressed.

A Gleeful Idea

Over the next few weeks, Pierce formulated a plan, and he shared it with his regional vice president. It was inspired by a practical joke he had played on a new employee a few months before. He had joked with this new person that the company had a glee club that met every Thursday night, and she was welcome to join. Of course, when she showed up, there was no glee club, but the idea of an imaginary glee club, a group that got together to join voices, stuck in Pierce's mind.

What if, Pierce imagined, the superintendents could get together and empower each other's voices? What if these amazing and smart professionals could learn and teach each other how to present their work to others? What if there were a glee club for Dynocon's superintendents?

Pierce knew he would never get this group to buy into this process without Dan Pricewater. Dan was a giant of a man and the most renowned project manager in Dynocon. Every superintendent in the company respected Dan. If Pierce could get Dan on board with the glee club idea, he had a chance. So, Pierce invited Dan to lunch and told him what he wanted to do and why. Pierce and Dan talked about how the only people who could genuinely build were the superintendents. Pierce spoke the truth when he said the client executives and project managers were little more than paper pushers without the superintendents. "However," Pierce told Dan, "the company is losing jobs because the superintendents cannot tell their stories, and they are great stories." Dan agreed.

The first meeting of the glee club was not a happy one even though it was held at the Skinny Armadillo Bar. Every single man, and they were all men, begrudged the time out of their day for something they considered a complete waste of time. As Pierce stood up to address this group for the first time, he knew most of the faces. Some of the men in this room knew and trusted him. He would have to build on that.

"Why are we here?" Pierce began. "This meeting is about winning." Pierce looked around the room. The music from the jukebox and the voices from the bar forced him to speak louder. "And we aren't winning—we are losing."

The superintendents all talked at once. Everyone had a reason for the recent bad run on project wins. Opinions poured out like beer from the bar taps. After a few minutes, Pierce spoke up,

> One of the reasons we are losing is because you guys walk into presentations like you are going to your execution. You can't seem to put two intelligent sentences together if your life depended on it. The owners look at you and say, I don't want that guy anywhere near my building.

Another round of complaining erupted from the men still dressed in their working clothes and steel-toed boots. "They don't pay us to talk, they pay us to build!" was the loudest refrain.

Dan's deep voice cracked through the collective din, "You won't have anything to build if we don't get some new jobs." The voices were silenced.

Then Pierce said,

> I have seen you solve the most complicated problems imaginable on your job sites. I have heard you explain complex systems and how they can be integrated. I have experienced your ability to sort out the most challenging logistics under the harshest time crunch. You can talk about what you do, but when you get in front of the owner, you get intimidated—for no reason. It doesn't matter where you sonsabitches come from or what education you have had. You can use your voice to create a feeling, to tell a story, and to create a connection. You can teach each other how.

Pierce went on,

> Many of you in this room have been building for more than three decades, and you know what you are doing. There is no reason you shouldn't be able to convincingly talk about your work anytime and anyplace without

preparation. This is simply about storytelling, and you have the ability to share your stories. Within six months, you will be able to just that.

The room continued to be quiet except for the background sounds from the bar until, finally, Dan said, "Sounds good to me. How do we start?"

Start Talking

Pierce grinned with anticipation as he watched the glee club members gather in the bar for the second meeting. The superintendents had been given homework. Each member of the glee club was to come prepared to tell a five-minute story about themselves, a story of how they got to where they were in life and in work today. They could tell it anyway they wanted, but it had to last five minutes. Pierce randomly selected the first speaker, Luis, to tell his story. Luis stood up a bit nervously but soon warmed to his subject, his father. Luis's father had been a carpenter, and Luis had learned all of his first lessons of construction from his father. He learned about hard work, about quality, about showing up, about planning, and about safety. His father had been a tough but caring teacher, and it broke his heart when he passed away ten years earlier. Luis closed by saying he carried the lessons of his father to the job site every day and tried to share them with others.

Only the background noises in the bar could be heard when Luis was done. The glee club members had listened in almost stunned silence. Then Dan said, "Well, alright!" and he started clapping, a sound soon joined by the claps of all Luis's peers. Then Pierce turned to the other superintendents and asked them what they thought about the speech. They just stared back at him. Pierce said, "This is part of the work. How did Luis do, and how could it have been better?" Pierce pointed to Harold, directly to his right, and said, "You start."

Harold was reluctant but finally said, "Well, I liked it. It pulled me in because he told a story about his dad. It was also the right amount of time." The second man to the right, Rick, spoke up, "At first he was nervous and rocked back and forth, but as he got into his story, he got better." More comments came easily about the details Luis shared and his willingness to share his emotions. Every superintendent was required to comment. Pierce observed the peer pressure was ramping up. The next talk, by Russell, was a bit rougher. He shared a story about getting in trouble in high school and how a friend's parent helped him get straightened out. He stopped before five minutes, and he had to find more to talk about. The comments from the glee club were harsher—"Your sentences were too short. You didn't share details. The story was good, but it could have been even more interesting"—Pierce realized that the glee club already understood what made a good speech. As each man presented, with a few exceptions, confidence in both speaking and commenting grew. It seemed almost everyone was engaged.

Pierce held his breath and then let it out slowly just before the third meeting of the glee club. This meeting is the real test, he thought. This meeting was going to be a tougher challenge. They were to think about a single area of interest outside of work that they had. It could be anything. Pierce told them that they would have to talk for five minutes on that subject, extemporaneously. When asked what that meant, Pierce said, "That means you just have to keep talking without preparation or notes for five minutes." The catch was that even though they would know the subject, Pierce would choose the specific topic. For example, if someone chose the subject of gardening, Pierce might ask them to talk about the best plants for late summer.

The first test was to see who showed up. Almost everyone, all fifteen, except one, showed up. Mark, one of the weakest speakers from the last meeting that took some of the harshest comments, decided to give up. Pierce wasn't surprised. Unlike the others who had accepted the comments with a teasing, "Oh, yeah?" Mark had shut down and seemed angry.

Today, Pierce selected names out of a jar to determine the speaking order. Ernie wanted to talk about car restoration. Pierce then asked him to talk about when it is desirable to upgrade or change engines and why. Ernie jumped into this topic with relish and exceptional detail. Pierce had to cut him off at five minutes even though Ernie had much more to say. The comments from the glee club members were enthusiastic and generous, although a couple of people thought there were a few too many details and engine specs included.

Hank wanted to speak about golf, so Pierce asked him to explain the difference in a swing with a pitching wedge and a driver. Hank had no problem filling five minutes with his description of the differences in the two strokes. He even figured out, on the fly, a clever way of organizing the information so the glee club members could remember the comparison.

The next speaker was Dan, who asked to talk about World War II. Pierce observed this giant of a man still wearing his red work shirt and overalls along with a cap settled over long red hair and beard. Pierce smiled to himself. Dan looked more like a farmer than a historian and quickly asked, "Which theater, Pacific or European?" "European" was the quick response. "So," Pierce challenged, "give me five minutes on Rommel's defense of Calais." Dan thought for a moment and began a short lecture that would have made a college professor proud, not to mention that it was downright impressive.

As the evening wore on, Pierce mentally scolded himself. He knew these men were smart and capable, but he had no idea just how interesting they were until today. He heard talks on dogs, cooking, cabinet making, furniture building, fishing, and several kinds of sports and exercise. One superintendent was writing a novel. Another was writing a book on hunting. Each talk was infused with in-depth knowledge of their subject matter and often deep passion. After all the speeches and comments were finished, Pierce rose and asked for everyone to be quiet.

> Each of you has spoken knowledgeably and easily for five minutes, without a script, about something they do in their spare time. And you did a Fucking Great Job. Don't tell me that you cannot speak without a script about something you have done nearly every day for the last thirty years of your working life. You most certainly can.

Every member of the glee club left that evening with a new sense of confidence.

As the meetings and the exercises continued, the glee club members learned about cadence, structure, and tone. They recited poems, tried interpretive readings, and practiced presentations. They chose their own subjects and were often challenged to try some new materials. At the Dynocon construction annual meeting where the entire company gathered, the glee club felt so comfortable, they offered to provide some entertainment. Superintendents recited poems individually and together to demonstrate their speaking abilities, and the glee club sang. The performance stunned the attendees, who began to see the superintendents differently. Superintendents were being asked to join presentations, and they were contributing positively. Pierce was asked to start glee club programs for other regions, and the president of the company insisted that project managers have a glee club of their own.

The New Guy

The following year, Sandy Archer was hired. Sandy was a legendary superintendent that had worked for a famous local construction company for thirty-five years until it had lost its patriarch and closed its doors. Sandy had run big projects of all kinds, including high rises, and he was one of the best superintendents in the business. Bringing Sandy to Dynocon was a coup. When Sandy was informed that he had to join the glee club, he was incensed. He wanted no part of a speaking club. He built buildings and left all that talking stuff to the suits. But Dan convinced Sandy to give the experience a chance, so Sandy arrived at his first glee club meeting cynical but willing. When it came time for Sandy to give his five-minute talk a few weeks later, Sandy wanted to talk about fishing boats. For years, he had been a tournament fisherman and had a complexion to testify to the long hours he had spent on the water. Unlike Dan, Sandy was small of stature, with long snow-white hair. The light yellow hair that had given him his nickname was long gone. Sandy had always been quiet spoken on the job site, but a natural intelligence and a hard glint in his eye had made it possible for him to lead his teams with great success.

As Sandy stood to talk about fishing boats, Pierce asked him to speak about different engine setups, including inboard and outboard motors. Sandy grinned, and without any evidence of nervousness started to talk. His voice was characteristically low, but he was generous with details and strategies. The comments from the other glee club members were varied but mostly complementary. Most wanted Sandy to speak up more with more expression. Sandy didn't seem upset by the criticism. Instead, he seemed to be processing something. When Pierce reiterated that Sandy could speak for five unscripted minutes on his hobby, and that he must undoubtedly be able to talk about his work, Sandy's head slowly nodded in acknowledgment.

Winning

Six weeks later, Sandy was asked to be part of a presentation for a major project. The project was a perfect fit for Dynocon and for Sandy, but the owner was unfamiliar with both. The presentation was held in a large conference room. All the presenters and the members of the owners' team were sitting around a large conference table. Sandy was sitting at the farthest end. He had a small part in the presentation, which he delivered competently and in his own words. It seemed everything was going fine, but there were no real moments of connection with the owner. The Q&A was halfway over.

Then a question about site safety was asked. Sandy realized this was his opportunity. He stood and leaned forward, demonstrating his interest and earnestness. Sandy smiled and breathed in deeply to fill his lungs and voice with air as he'd been taught. He was satisfied with the confident tone that emerged as he told the owner how important this question was and how important it was to him. Yes, he organized his job sites for efficiency and project success, but no project is successful if someone didn't go home to their family. Then he described Dynocon's safety program and the elements he felt made it unique. He also talked about the way he spoke to his teams about safety every day.

When Sandy finished, the entire room was engaged in his topic. The owner asked several follow-up questions, directly to Sandy, which Sandy answered with confidence and knowledge. When the interview ended, the owner shook Sandy's hand warmly. "Thank you for coming," he said, "I can tell you've been doing this a long time." "Yes sir, I'm your man," said Sandy, and the owner smiled.

Dynocon was notified a day later they had been chosen for the project—provided Sandy was guaranteed to be the superintendent. Sandy never, ever missed a meeting of the glee club.

Index

Note: Figures are indicated by *italics* in this index.